D0409770

Luminescence Techniques in Solid-State Polymer Research

edited by

Lev Zlatkevich
VIBO Research, Inc.
Pennsauken, New Jersey

TXD 27/7/89

MARCEL DEKKER, INC.   New York and Basel

Library of Congress Cataloging-in-Publication Data

Luminescence techniques in solid state polymer research / edited by
   Lev Zlatkevich.
        p.   cm.
      Includes index.
      ISBN 0-8247-8045-0
      1. Polymers and polymerization.   2. Luminescence.   3. Solid state
   chemistry.   I. Zlatkevich, L. (Lev)
   QD381.8L85   1989
   547.7--dc19   88-26745
                                                                    CIP

This book was printed on acid-free paper.

MARCEL DEKKER, INC.
270 Madison Avenue, New York, New York   10016

Current printing (last digit):
10 9 8 7 6 5 4 3 2 1

PRINTED IN THE UNITED STATES OF AMERICA

Preface

The sensitivity of luminescence measurements has rendered them of great use in various branches of polymer research. This technique often provides information unobtainable by other means and has been used in many areas of basic and applied research, production, and application.

Previous books dealing with polymer luminescence have tended to emphasize studies of polymers in solution. This book was designed with the objective of covering numerous important applications of luminescence to polymers in solid state and is the first attempt to present this topic as a single subject.

Dr. George opens the volume with a discussion of various processes that constitute polymer luminescence and indicates in general terms the solid-state properties that may be studied in each case.

In Chapter 2, Drs. Allen and Owen review various applications of luminescence (fluorescence and phosphorescence) spectroscopy in polymer analysis and provide a number of prominent and representative examples of the more practicable uses of the technique.

The focus of Chapter 3, written by Dr. George, is on chemiluminescence of polymers at nearly ambient conditions. Various methods for observing chemiluminescence under nonstationary conditions are described and an up-to-date assessment is presented.

In Chapter 4, I discuss chemiluminescence experiments designed primarily for evaluation of materials' thermal oxidative stability. A new method of obtaining both major parameters of autoxidation—induction time and oxidation rate—is discussed in detail and applied

to the study of uninhibited and inhibited oxidations as well as materials' long-term stability.

This is followed by Drs. Allen and Owen's discussion on luminescence studies of photooxidation in various polymers with the emphasis
on similarities between different systems.

Next, a survey of studies on stress-induced chemiluminescence
is presented by Drs. Monaco and Richardson. Although qualitative
in nature, the results obtained so far suggest that the technique can
be of value in predicting the probability of failure of a stressed material in a given environment.

The concluding chapter, written by myself, reviews the application of radiothermoluminescence to the study of temperature transitions in polymers.

The book describes the most important practical aspects of luminescence relevant to the study of polymers in solid state. Many
recent developments are included, and some of the methods and techniques presented may serve as models for the reader who wants to
apply them to some other related problem in research, production,
and quality control.

The intention of this volume is to narrow the gap between theory
and practice. It is aimed primarily at research workers in industrial
and governmental laboratories and at polymer technologists and aca-
demics. It will also be useful for postgraduate and advanced under-
graduate students in polymer sciences.

Lev Zlatkevich

Contents

Contents

Contributors

Norman S. Allen   Department of Chemistry, John Dalton Faculty of Technology, Manchester Polytechnic, Manchester, England

Graeme A. George   Department of Chemistry, Queensland Institute of Technology, Brisbane, Australia

Suzanne B. Monaco   Chemistry and Materials Science Department, Lawrence Livermore National Laboratory, Livermore, California

Eryl D. Owen   Department of Chemistry, University College, Cardiff, South Wales

Jeffery H. Richardson   Chemistry and Materials Science Department, Lawrence Livermore National Laboratory, Livermore, California

Lev Zlatkevich   VIBO Research, Inc., Pennsauken, New Jersey

1

Luminescence in the Solid State
General Requirements and Mechanisms

GRAEME A. GEORGE   Queensland Institute of Technology, Brisbane, Australia

‹ab p.46

I. INTRODUCTION

In the broadest sense, polymer luminescence is the light emitted by a polymer after it has been excited in some way.   The excitation process may involve the absorption of high-energy radiation, UV-visible radiation, the excess energy of a chemical reaction or a physical process such as charge recombination occurring in the solid polymer.   The starting point in all studies of luminescence is the measurement of the emission spectrum and the decay properties of the luminescence.   Much of our present understanding of the luminescence from solid polymers has resulted from the study of the properties of small organic molecules.   The photoluminescence of organic molecules, often at cryogenic temperatures, has provided information about the symmetry of the molecule, its electronic excited states, and the pathways for dissipation of energy, including the interaction with other molecules [1,2].

Similar information can also be obtained from organic polymers and for many years the luminescence properties of polymers have been measured and related to those of the small-molecule analogue: the polymer repeat unit [3].   In many of the fundamental studies of polymer luminescence, the spectral properties are measured in solution in order to avoid the perturbing effects of nearest-neighbor interactions and thus enable the study of polymer conformation and chain dynamics [4].   Time-resolved fluorescence spectroscopy has been most powerful in providing this information [5].

1

The solid state properties of polymers are of great practical significance and luminescence spectroscopy is playing an important role in such diverse areas as chemical and physical identification; morphology and orientation; thermal properties and transitions; polyblend compatibility; transport properties including permeability; degradation; stabilization; deformation; and fracture. In addition to studies of the intrinsic luminescence of the polymer, luminescence from impurities and additives may be readily observed in the solid polymer matrix to enable the study of the excited state properties of the additive, and also to probe the structure of the polymer.

All of these applications will be addressed in this book. The purpose of this chapter is to introduce the various processes that constitute polymer luminescence and indicate in general terms the solid state properties that may be studied in each case. The process of light absorption and emission will initially be considered for small organic molecules. This will then be extended to consider the luminescence properties in the solid state and in particular when the molecules form part of the chain of a polymer.

II. PHOTOPHYSICS OF SMALL ORGANIC MOLECULES

The available pathways of excitation and deexcitation of a small organic molecule (e.g., a polyacene such as naphthalene) may be most conveniently described by a state diagram [6] as shown in Figure 1. In this diagram, the energies of the possible excited states of the molecule are referenced to its lowest energy or ground state. In almost all cases this is a singlet state S_0 in which the spin states of the electrons are paired.

Higher energy singlet states S_1, S_2, and so on, correspond to promotion of an electron to an unoccupied molecular orbital. Under normal conditions of temperature and pressure, the large energy gap ensures that these levels are initially unpopulated. In Figure 1, sufficient energy has been provided by a quantum of radiation to promote an electron to the first excited singlet state S_1. It is noted that there is no change in spin state in this process and this corresponds to an $S_1 \leftarrow S_0$ electronic transition. Because of the coupling of vibrational and electronic wave functions, vibrational quanta may also be excited in this absorption process. Thus a measured absorption spectrum will show a vibrational progression built on the pure electronic transition. From the Franck—Condon principle, the most intense absorption may correspond to the excitation of a higher vibrational level of S_1 since this configuration is closer to the geometry of the ground state S_0. The vibrational sublevels are shown for each electronic state.

The process of deexcitation of the molecule thus consists of following the various routes available to return the molecule to the ground state S_0. These are the photophysical paths and it is assumed that no photochemical reaction such as dissociation is occurring from any excited state. Which of the particular paths shown in Figure 1 will be followed depends on the values of the rate constants for each. In the first instance the molecule is considered to be isolated so that only unimolecular processes occur. A simple kinetic scheme can then be set up under photostationary conditions to express the rates of these competing processes.

A. Unimolecular Processes

The unimolecular processes and their rates are shown in Table 1. If a higher vibrational level S_1^V is initially excited, the first process of energy loss is internal conversion to the zero vibrational level S_1. The rate constant for this reaction is very large ($\sim 10^{13}$ sec^{-1}) on the order of the molecular vibrational frequency, so that the state S_1 is populated regardless of the vibrational level excited. From Figure 1 it can be seen that the available pathways from S_1 to S_0 are the emission of light as fluorescence, $S_1 \rightarrow S_0$, radiationless decay by internal conversion (which because of the large energy gap between S_1 and S_0 will be much slower than between S_1^V and S_1) or intersystem crossing to the triplet manifold. The triplet

TABLE 1   Photophysical Processes and Their Rates

Process	Rate
A. Unimolecular	
Absorption: $S_1^V \leftarrow S_0$	I_a
Internal conversion: $S_1^V \rightarrow S_1$	$k_{IC}[S_1^V]$
Internal conversion: $S_1 \rightarrow S_0$	$k_{IC}[S_1]$
Intersystem crossing: $S_1 \rightarrow T_1^V$	$k_{ISC}[S_1]$
Fluorescence: $S_1 \rightarrow S_0$	$k_F[S_1]$
Phosphorescence: $T_1 \rightarrow S_0$	$k_p[T_1]$
Intersystem crossing: $T_1 \rightarrow S_0$	$k_{ISC}'[T_1]$
B. Bimolecular	
Quenching: $S_1 + Q \rightarrow S_0 + Q^*$	$k_Q[S_1][Q]$
$T_1 + Q \rightarrow T_0 + Q^*$	$k_Q'[T_1][Q]$
Complex formation: $S_1 + S_0 \underset{k_{MD}}{\overset{k_{DM}}{\rightleftarrows}} (S_1 S_0)^* \overset{k_{FD}}{\rightarrow} S_0 + S_0 + h\nu$	
Triplet annihilation: $T_1 + T_1 \rightarrow S_1 + S_0$	$k_b[T_1]^2$

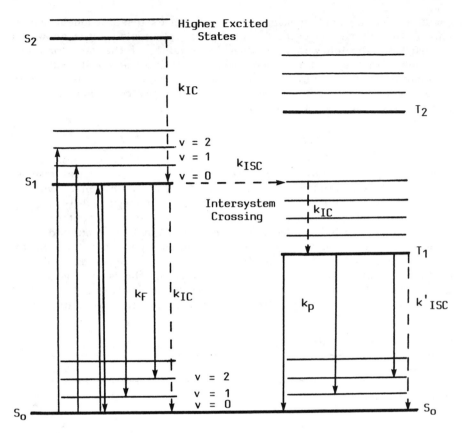

FIGURE 1 Absorption, fluorescence, and phosphorescence spectra
of a small organic molecule and the corresponding energy state
diagram showing radiative (→) and radiationless (--→) processes and
their rate constants. Only unimolecular processes are shown.

energy levels T_n correspond to the spin states of the electron being
parallel rather than paired. While direct population of the triplet
state by absorption of radiation is spin-forbidden, it may be populated by intersystem crossing from S_1 since, once in the excited
state and occupying separate molecular orbitals, there are no spin
constraints. Because of electron repulsion between parallel spin
states, the energies of the triplet state are lower than those of the
corresponding singlet states. The magnitude of the energy difference
between S_1 and T_1 depends on orbital overlap and is greater in
$\pi\pi^*$ excited states than in $n\pi^*$ excited states.

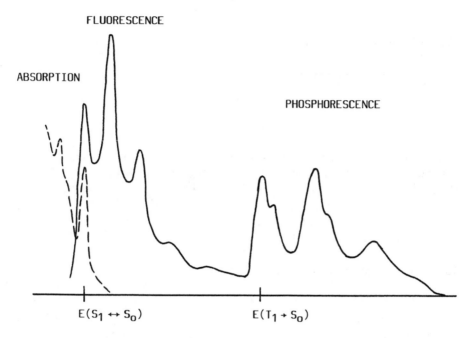

FIGURE 1   Continued

Once T_1 is populated, energy may be lost either by the radiative process $T_1 \rightarrow S_0$ (which is shown in Figure 1 as phosphorescence) or by the radiationless intersystem crossing to the ground state. Again, any possible photochemical reactions from T_1 are not presently considered.

If steady-state excitation is occurring, $d[S_1]/dt = 0$, that is, absorption is constant with time and the system will be in a stationary state where the rate of absorption will equal the rate of deexcitation. From Table 1 this gives

$$I_a = (k_F + k_{ISC} + k_{IC})\ [S_1] \tag{1}$$

The relevant fluorescence parameters may now be determined.   These are

(1)   Fluorescence quantum yield

$$\phi_F = \frac{k_F[S_1]}{I_a}$$

$$= \frac{k_F}{k_F + k_{ISC} + k_{IC}} \tag{2}$$

and

(2) Fluorescence lifetime

$$\tau_F = \frac{1}{k_F + k_{ISC} + k_{IC}}$$

$$= \frac{\phi_F}{k_F} \tag{3}$$

Applying the stationary state approximation to the triplet state and setting $d[T_1]/dt = 0$, from Table 1 we obtain $k_{ISC}[S_1] = (k_p + k'_{ISC})[T_1]$. Substituting from Equation (1),

$$[T_1] = \frac{I_a k_{ISC}}{(k_F + k_{ISC} + k_{IC})(k_p + k'_{ISC})}$$

The relevant phosphorescence properties are then given by

(1) Phosphorescence quantum yield

$$\phi_p = \frac{k_p[T_1]}{I_a} = \left(\frac{k_p}{[k_p + k'_{ISC}]}\right)\left(\frac{k_{ISC}}{k_f + k_{ISC} + k_{IC}}\right)$$

$$= \phi'_p \, \phi_{ISC} \tag{4}$$

where
ϕ'_p = quantum efficiency of phosphorescence
ϕ_{ISC} = quantum yield of triplet state formation

(2) Phosphorescence lifetime

$$\tau_p = \frac{1}{k_p + k'_{ISC}} \tag{5}$$

It is noted from Figure 1 that the fluorescence and phosphorescence spectra will also consist of a series of bands corresponding to transitions from S_1 (v = 0) and T_1 (v = 0) to vibrational levels of the ground state. The energy separation of these bands should be the same as those obtained by vibrational spectroscopy of the ground state. The fluorescence and phosphorescence lifetimes may be determined by measuring the decay of the emission after ceasing irradiation. First order kinetics should be obeyed in an isolated molecule.

The values of the measured emission parameters of luminescence yield and lifetime will depend only on the values of the first order rate constants if only unimolecular pathways are followed. Typical values for a chromophore with $\pi\pi^*$ excited states (e.g., naphthalene) and $n\pi^*$ excited states (e.g., benzophenone) are given in Table 2. For molecules containing hetero atoms (and thus $n\pi^*$ excited states), k_{ISC} is increased and it is generally found that these have a larger ϕ_p, small ϕ_F, and larger k_p (i.e., a very short phosphorescence lifetime on the order of msec compared to several seconds for the $\pi\pi^*$ excited states). It is noted that with both $n\pi^*$ and $\pi\pi^*$ states the phosphorescence lifetime is many orders of magnitude longer than the fluorescence lifetime. This reflects the spin-forbidden nature of the $T_1 \leftrightarrow S_0$ transition.

B. Bimolecular Processes

While it is possible to examine the photophysics of isolated individual molecules in dilute rigid solutions to avoid bimolecular reactions, in polymers the close spacing of repeat units ensures that an excited molecule is going to interact with its nearest neighbors. If this interaction can occur within the excited state lifetime, then alternative, bimolecular pathways for loss of excitation energy must be considered.

Kinetically this can be treated by adding a number of possible bimolecular rates for the singlet and triplet states to Table 1. While several different bimolecular processes (such as enhanced intersystem crossing) have been recognized, the most important processes are excited state quenching and complex formation. Among quenching

TABLE 2 Photophysical Parameters for Napthalene and Benzophenone

	Naphthalene	Benzophenone
Singlet:	$^1\pi\pi^*$ E = 386 kJ/mol	$^1n\pi^*$ E = 318 kJ/mol
	$k_F \sim 10^7$ sec^{-1}	10^6 sec^{-1}
	ϕ_F 0.55	0.0
Triplet:	$^3\pi\pi^*$ E = 255 kJ/mol	$^3n\pi^*$ E = 291 kJ/mol
	$k_{ISC} \sim 10^7$ sec^{-1}	5×10^{10} sec^{-1}
	k_p 0.04 sec^{-1}	160 sec^{-1}
	ϕ_p 0.05	0.84

reactions, energy transfer from the excited molecule to another
molecule with lower lying energy levels of the correct multiplicity
is most common. Many bimolecular reactions of excited states become
diffusion-controlled because of the very high value of the rate
constant for energy transfer and quenching, i.e., k_Q becomes
replaced by the diffusion-controlled rate constant which is dependent
on the viscosity of the medium. If the quencher is a small molecule
such as oxygen, k_Q is dependent on the diffusion coefficient of the
gas in the material.

Considering just singlet states, the expression for the fluorescence quantum yield [Equation (2)] now becomes

$$\phi_F = \frac{k_F}{k_F + k_{ISC} + k_{IC} + k_Q[Q]} \tag{6}$$

If ϕ_F^0 is used to represent the original fluorescence quantum yield
in the absence of quencher and τ_F^0 the original lifetime, then
combining Equations (2) and (6):

$$\frac{\phi_F^0}{\phi_F} = 1 + k_Q \tau_F^0 [Q] \tag{7}$$

which is the familiar Stern-Volmer Equation [6].

Treatment of the triplet states generates a similar expression
for the quenching of phosphorescence and as can be seen from Table
2, since τ_p^0 is typically 10^6 τ_F^0, the triplet state is much more
susceptible to quenching than the singlet state. Since oxygen is
paramagnetic, it is a ubiquitous quencher of triplets, which is why
the phosphorescence quantum yield is negligibly small in nondegassed
solution. Consequently, the experimental observation of phosphorescence has traditionally required low temperatures and the examination of samples in vacuo to reduce both k_Q (which in fluid solution
will be the diffusion-controlled rate constant) and the value of [Q].
However, as will be discussed later, in some solid polymers it is
possible to observe phosphorescence at room temperature in the
presence of air because of the low solubility and diffusion coefficient
of oxygen.

In solutions of small molecules, complex formation and triplet
annihilation as shown in Table 1B would be expected to occur at
high local concentrations of chromophores or at very high light
intensities. In polymers, whether in solution or the solid state, in
which the chromophore occurs regularly as the repeat unit along the
polymer backbone, the local concentration will be high even in dilute
liquid or solid solution. The photophysical process of complex formation, in particular excited dimers or "excimers," is thus of great
importance in understanding the luminescence properties of polymers [7].

1. Excimer Formation

An excimer is an excited state complex formed between an excited molecule and an unexcited molecule. Experimentally this often appears as a change in the fluorescence spectrum as the structured emission (Figure 1) is replaced by a broad, structureless spectrum of lower energy (excimer fluorescence). As shown in Figure 2 for pyrene, the relative intensity of the excimer emission compared to the normal fluorescence increases with concentration over the range 10^{-3} mol/dm^3 to 10^{-1} mol/dm^3. No comparable change is observed in the absorption spectrum indicating that the excimer state is not a ground state dimer but exists only in the excited state.

This can be seen from the potential energy diagram in Figure 2. Only when the planes of the aromatic molecules adopt a sandwich configuration with a separation of $\sim$3.3 Å will a stable excimer be formed as indicated by the minimum in the potential energy curve. It is noted that the comparable ground state is dissociative and at separations >4 Å emission from the free molecule is observed. The kinetic treatment of excimer formation and decay from Table 1B gives the ratio of excimer emission quantum yield ϕ_E to free molecule quantum yield ϕ_F as

$$\frac{\phi_E}{\phi_F} = \frac{k_{FD} \cdot k_{DM} \cdot [S_0]}{k_F(k_{MD} + k_{FD} + k_{ID})} \tag{8}$$

This shows the excimer emission to increase with concentration of the chromophore in solution $[S_0]$. k_{ID} is the rate constant for radiationless decay of the excimer to the ground state. The time dependence of excimer fluorescence [2] has been used to investigate the dynamics of excimer formation and decay to complement steady-state measurements using Equation (8). This has been of most importance in the dynamics of macromolecules in solution [4].

2. Exciplexes

A complex analogous to the excimer may also be formed between the excited molecule and a chemically different ground state molecule. This "exciplex" may dissociate radiatively to produce a structureless emission spectrum similar to excimer emission or, most commonly, decay radiationlessly [6]. It has been suggested that most bimolecular quenching processes described by the Stern-Volmer equation occur through the formation of an encounter complex between the excited state and the quenching molecule with the formation of an exciplex. In solution the rate of formation of the exciplex will be the diffusion-controlled rate. Thus, provided the rate constant for formation of the ground state products is greater than the rate constant for the reverse reaction of dissociation of the exciplex to

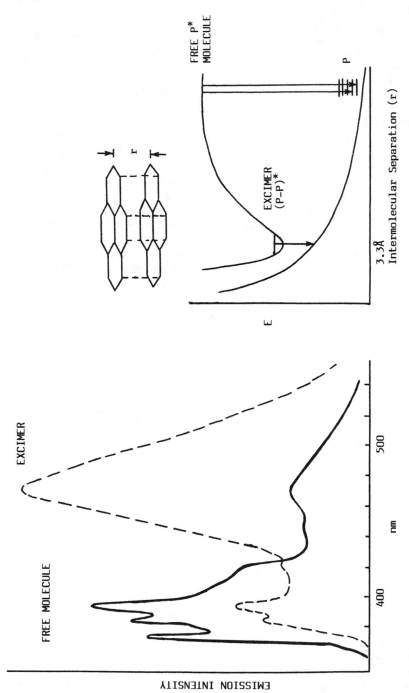

FIGURE 2 Excimer formation in pyrene. The fluorescence spectrum of pyrene in ethanol at 3×10^{-4} M (——) and 3×10^{-3} M (- - -) shows the appearance of a lower energy structureless emission that is explained by the potential energy diagram with excimer formation at 3.3 Å. (Adapted from Ref. 6, p. 104.)

the excited state, the quenching rate constant will be the diffusion-controlled rate constant:

$$k_Q = k_d = \frac{8RT}{3\eta}$$ (9)

from the Debye Equation [6].

Many of the diffusion-controlled quenching processes may be recognized as involving an exciplex which then dissociates to the ground state without emission of radiation. The following examples are typical.

1. The enhancement of triplet state yield when a heavy atom is present at high concentration in solution may be considered as due to the formation of a singlet exciplex. The increased spin-orbit coupling, due to the atom of high atomic number, enhances the intersystem crossing to the triplet exciplex, which then dissociates. If M^* is the excited molecule and Q the quenching molecule containing a heavy atom,

$$^1M^* + Q \rightarrow {}^1(MQ)^* \rightarrow {}^3(MQ)^* \rightarrow {}^3M^* + Q$$

Examples in the photophysics of aromatic hydrocarbons are the enhanced phosphorescence quantum yield and shortened triplet lifetime when halogen atoms are present.

2. The quenching of singlet excited states by paramagnetic molecules such as oxygen and nitric oxide, while it may involve energy transfer, could also involve exciplex formation and enhanced intersystem crossing.

$$^1M^* + {}^3O_2 \rightarrow {}^3(MO_2)^* \rightarrow {}^3M^* + {}^3O_2$$

3. Electron transfer quenching, such as amine quenching of hydrocarbon fluorescence in polar solvents to produce solvated ion pairs, may occur through an exciplex

$$^1M^* + Q \rightarrow {}^1(M^*Q) \rightarrow M_s^+ \cdot Q_s^-.$$ solvated ion pair

3. Triplet—Triplet Annihilation

At high intensities of absorbed radiation or with a molecule having a long excited state lifetime, it is possible to build up quite high concentrations of excited states. This is most common with the triplet excited state and it is well known that $T_n \leftarrow T_1$ absorption spectra may be observed after flash excitation [1,2]. Another triplet state phenomenon under these conditions is the bimolecular reaction of two excited molecules to produce an excited singlet state and a ground state molecule as shown in Table 1B. This process may also take place through the formation of an exciplex as the excited state intermediate as discussed in the previous section.

$$^3M^* + {}^3M^* \rightarrow {}^1(MM)^* \rightarrow {}^1M + {}^1M^*$$

If the molecule has a nonzero fluorescence quantum yield, then emission will be observed spectrally identical to the fluorescence, but with a lifetime close to that of the triplet state. This emission is termed "P-type" <u>delayed fluorescence</u> to distinguish it from a delayed emission that arises by thermally assisted repopulation of the singlet state from T_1. This occurs when the energy gap S_1-T_1 is on the order of kT and intersystem crossing from T_1 back to S_1 is possible. Again the emission will be identical to fluorescence and longer lived and is termed "E-type" delayed fluorescence. It can be readily distinguished from P-type emission since the emission intensity is proportional to the exciting light intensity and the delayed fluorescence lifetime is the same as the triplet lifetime. In contrast, the P-type emission intensity is proportional to the square of the exciting light intensity and the lifetime is half that of the triplet state lifetime. This phenomenon is of importance in studies of solid polymers where annihilation can only occur between triplet excited molecules which are formed adjacent to one another, since they cannot diffuse together in the solid matrix within the excited state lifetime. The observation of delayed fluorescence at low excitation intensity is taken as evidence for the delocalization of excitation energy over several repeat units of the polymer leading to migration of the energy, and thus triplet—triplet annihilation. This is discussed in the next section as one of the distinct features of the high local concentration of chromophores always present in the polymer.

III. POLYMER PHOTOPHYSICS

The process of the absorption and emission of radiation by a chromophore on the backbone of a polymer chain may be considered to be similar to the process described by Figure 1 and discussed in the previous section, except that the high local concentration means that the spectral properties may be affected by the position of the chromophore on the chain, the number and type of other chromophores that are present, and the physical state of the polymer. We can also distinguish the situation of an idealized polymer chain containing the chromophore (either as the repeat unit or as an isolated center) dissolved in a good solvent, from the real structure in the solid state which is highly defective. These defects may be chemical in nature such as head-to-head linkages, oxidation products, and adventitious impurities or physical defects such as stereo irregularity, free-volume fluctuations, and, in a semicrystalline polymer, the wide range of morphological defects that lead to variations in the micro-environment and mobility of the polymer chain.

For an idealized, isolated chain, Holden and Guillet [3] summarized the type of chromophore distribution that may be expected in synthetic polymers and the polymer properties that affect the luminescence. Table 3 illustrates some of these types and uses the convention of describing the polymers as either type A or type B.

Type A: The chromophores are essentially isolated and the emission
         properties should more closely resemble those of the free molecule.
         The chromophore may be a deliberate label on the chain or it
         may be an adventitious impurity such as an oxidation product.
Type B: The chromophores occur regularly and at a high local
         concentration, so that the spectral properties are affected by
         the resulting intermolecular interactions. Examples include both
         homopolymers and copolymers.

In studies of energy migration and trapping in polymers, most interest has centered on type B polymers because of the high, controlled chromophore concentration. However, in many commercial polymers, type A often represents the situation of a solid polymer with impurity groups distributed at random as a result of synthesis or processing. The luminescence properties of the polymer may then be dominated by these isolated chromophores if they have low-lying electronic energy levels. To understand the emission spectra, it is necessary to consider the processes of electronic energy transfer that can occur in the solid state.

A. Energy Migration in the Organic Solid State

The development of the understanding of energy migration in polymers has built on the studies of the same chromophores in molecular crystals. The polymer chain, while regarded by some as a one-dimensional molecular crystal [2], cannot be regarded as having the necessary symmetry elements to be treated formally in this way. At best it may be treated as a highly defective, one-dimensional array of chromophores to which the general principles of energy migration in the organic solid state may be applied.

Energy delocalization in organic molecular crystals results from the coupling of the electronic excited state of a chromophore with an adjacent identical unexcited molecule [8]. As a result of the intermolecular interactions between the molecule in the crystal and its immediate neighbors (which for singlet excited molecules will be Coulombic in nature), the excitation can no longer be considered as occurring on an isolated molecule but rather over many identical chromophores. On a single-molecule picture this means that the time

TABLE 3 Distribution of a Chromophore Molecule X (e.g., aromatic group, carbonyl group) in an Idealized, Extended Chain Polymer

TYPE A

End of Chain

Side-Chain

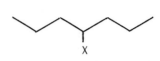

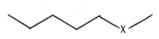

X

In-Chain

X

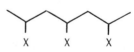

TYPE B

Homopolymer In-Chain

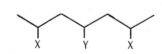

Homopolymer Side-Chain

X   X   X

Copolymer (Alternating)

X   Y   X

Copolymer (Random)

X   X   Y   X

for the excitation energy to migrate to a neighboring molecule is comparable with the time for the absorption process. This process may be regarded as the motion of a spinless particle—the exciton— through the crystal lattice. For singlet exciton migration the mechanism of exciton transfer requires dipole—dipole coupling between the molecules and results in a shift in the absorption spectrum of the crystal and a splitting of an absorption band into two or more oppositely polarized components [8].

The magnitude of this interaction energy I is given by an equation of the form

$$I = \frac{e^2}{R^3} M^2 f(\theta) \tag{10}$$

where

e = electron charge
R = intermolecular separation
M = transition moment for the electronic transition
$f(\theta)$ = a function of the angles made by the transition moment with the crystallographic axes

The important feature of Equation (10) is that the interaction potential I depends on the intensity of the electronic transition (i.e., the molar absorptivity for the $S_1 \leftarrow S_0$ transition) and varies as the inverse third power of the intermolecular separation. The extent of the delocalization depends on the perfection of the lattice and is limited by interactions with lattice vibrations and defects [9]. In the limit of high scattering, the exciton migration may be best regarded as a random walk or diffusive process. In a molecular crystal of naphthalene ($R \sim 4$ Å), the singlet migration distance may span several thousand chromophores.

The above treatment applies only to singlet excited states since the major contribution to I arises from Coulombic terms. For triplet states these terms will be zero in the absence of significant spin-orbit coupling and the interaction energy can arise only from electron exchange integrals. Thus the interaction potential leading to delocalization of triplet excitation in a molecular crystal is very small (5 to 10 cm^{-1}) compared to the singlet state (10^2 to 10^4 cm^{-1}). As a result, the rate of triplet energy migration will also be much smaller. However, against this the radiative lifetime of the triplet state of an aromatic hydrocarbon such as naphthalene is up to 10^9 times that of the singlet state (Table 2). This means that the number of energy transfer steps per excited state lifetime may be greater for the triplet exciton than the singlet exciton. Again, the migration of the triplet exciton may be regarded as a diffusive random walk process.

The interaction potential for triplet states has neglected spin-orbit coupling. In molecules containing heteroatoms (e.g., carbonyls) and heavy atoms such as halogens, the spin-orbit coupling introduces significant singlet state character into the triplet state. There will now be a coulombic component [Equation (10)] in the interaction potential, which may be $\sim 10^2 I$. This allows triplet energy migration in systems which may have small electron exchange integrals because of symmetry or large intermolecular separation.

The migration of an exciton through the crystal lattice will continue until

1. It interacts with either a physical or chemical defect in the lattice and is trapped. Energy may be lost by either radiationless decay or emission from the trapped state. If the molecules at the trap site are in the correct geometrical configuration to form an excimer, then structureless excimer emission of the type shown in Figure 2 will be observed.
2. If it is a triplet exciton, it interacts with another triplet exciton and mutual annihilation occurs to produce delayed fluorescence by the mechanism described in the previous section. In molecular crystals such as anthracene this process is so efficient that the quantum yield of phosphorescence from the pure crystal is very low as most triplets annihilate within the excited state lifetime. In this case the dynamics of the triplet exciton have been probed by studying the delayed fluorescence.
3. It undergoes spontaneous emission to the ground state. This requires that neither of the processes 1 and 2 have occurred within the excited state lifetime.

The study of the luminescence properties such as excimer emission and delayed fluorescence have provided direct evidence for triplet energy migration. Similarly, the observation that only 1 part per million of an impurity molecule is required to change the emission spectrum to that of the impurity is an indication of facile energy migration and trapping. These luminescence phenomena are also observed in polymers and it is this similarity that has led to the analogy of an idealized polymer chain as a one-dimensional molecular crystal. In the solid state the problems of stereo regularity and chain conformation fluctuations lead to the concept of a polymer as a highly defective structure in which only limited delocalization of excitation energy can be expected.

B. Energy Migration and Trapping in Polymers

With the above limitations of the symmetry requirements in mind, the concepts of singlet and triplet energy migration and trapping may be applied to the type B polymer such as poly(2-vinylnaphthalene) [10—12]. The energy migration may be viewed as a one-dimensional random walk with a diffusion coefficient D_E (m^2/sec). If τ_E is the excited state lifetime of the chromophore and τ_h is the hopping time for the energy (i.e., the inverse of the energy transfer rate constant), then the mean free path for energy migration $\bar{l}$ is given by

$$\bar{l} = (2D_E\tau_E)^{1/2} = R\left(\frac{\tau_E}{\tau_h}\right)^{1/2} \tag{11}$$

Thus the average range for energy transfer is given by

$$\frac{\bar{l}}{R} = \left(\frac{\tau_E}{\tau_h}\right)^{1/2} \tag{12}$$

Values for vinyl aromatic polymers such as poly(2-vinylnaphthalene) are reported to be $\bar{l} \approx 11.5$ nm, $\tau_h \approx 5 \times 10^{-11}$ sec, $D_E \cong 1.5 \times 10^{-7}$ m^2/sec. This would suggest a range of about 30 chromophores. Semerak and Frank calculated an upper limit of 300 chromophores by using parameters from crystalline naphthalene [7].

The fate of the excitation is either trapping of the energy (at an impurity- or excimer-forming site), annihilation by encountering another excited state (if triplets are considered, producing delayed fluorescence), or annihilation by molecular emission (often referred to as "monomer" emission). Some of these processes are shown schematically in Figure 3. The actual spectra obtained and the information regarding the polymer structure will be discussed in Section IV.

The process of excimer formation in polymers has been the subject of close investigation [7]. Two different mechanisms for excimer formation have been considered. The first is that there exists in the polymer a number of excimer-forming sites in which a pair of chromophores is already in the correct planar arrangement for excimer formation. When one of these is excited in the process of energy migration, the energy is instantly trapped and excimer emission is observed.

Alternatively, chain rotation may occur within the excited state lifetime to allow dynamic formation of the excimer and thus trapping of the energy. Which of these processes is favored seems to depend on the physical state of the polymer [7,10]. For example, in dilute solution, singlet excimer formation is not observed at low temperature, while it still occurs from solid films. It has been reported that the fraction of excimer-forming sites in polystyrene film is as high as 40% [7].

It should also be noted that the excimer does not have to form between adjacent chromophores along the polymer chain. It is possible through chain interpenetration that excimers form from chromophores on different chains or by entanglements through non-adjacent chromophores on the same chain. The quality of the solvent

A : ENERGY MIGRATION TO FORM EXCIMER

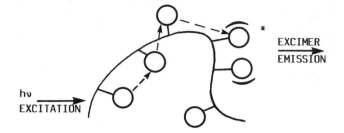

B : TRIPLET MIGRATION AND ANNIHILATION

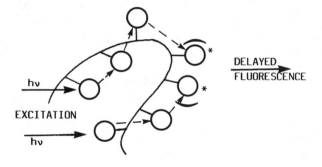

C : ENERGY MIGRATION TO TRAP

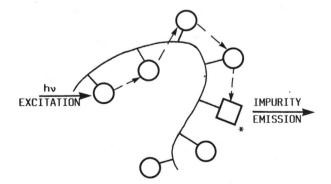

FIGURE 3 Typical energy migration processes in a vinyl aromatic polymer and the resulting emission phenomena. Processes A and C may occur in either the triplet or singlet state.

will have a bearing on the excimer formation probability for polymers in dilute solution since the coil size will change. This will also apply in the solid state, and the interpenetration of polymer networks in miscible blends will alter the excimer formation efficiency.

Energy transfer processes in the polymer containing more than one type of chromophore are also sensitive to the extent of chain interpenetration. These are discussed in the next section.

C.   Energy Transfer in the Solid State

1.   Singlet States

It has been noted that the excited state interaction that led to migration of the singlet excitation energy over identical chromophores in a molecular crystal or along the polymer backbone resulted from Coulombic interactions [Equation (10)]. This resulted in strong coupling of the excited states and typical energy migration distances of tens of nanometers. Singlet excitation energy may also be delocalized in a polymer by coupling between nonidentical chromophores.

The process of excited state quenching in solution described by the Stern-Volmer Equation (7) occurs through the formation of an encounter complex or exciplex at the diffusion-controlled rate between the excited chromophore and the quencher molecule, which has lower lying energy levels of the correct multiplicity (Section II.B). In this mechanism it would be expected that quenching could not occur in the solid state (where both the excited molecule and quencher are unable to diffuse together to form the encounter complex) unless very high concentrations of quencher are used, so that the molecules are only separated by the interaction diameter. It is found, however, that energy transfer in rigid systems can occur over distances substantially greater than this [6].

For singlet energy transfer, Forster showed that the interaction energy also arises from Coulombic (dipole-dipole) forces between the excited molecule (donor D^*) and the quenching molecule (acceptor A).

$$D^* + A \xrightarrow{k_{DA}} D + A^* \tag{13}$$

The rate constant k_{DA} for this reaction will be sensitive to intermolecular separation and the spectral properties of both donor and acceptor.

$$k_{DA} = \frac{K k_{FD}}{R_{DA}^6} S_{DA} \tag{14}$$

where

k_{FD} = singlet state radiative rate constant of the donor

R_{DA} = intermolecular separation

K = a constant that is a function of refractive index of the medium and the relative orientation of D^* and A

S_{DA} = a spectral overlap integral

S_{DA} depends on the extent of overlap between the fluorescence emission spectrum of D^*, F_D, and the abosprtion spectrum of A, ε_A, on a wavenumber scale.

$$S_{DA} = \int_0^\infty F_D \cdot \varepsilon_A \frac{d\bar{\nu}}{\bar{\nu}^{-4}} \tag{15}$$

The energetic and spectral requirements for energy transfer by the Forster mechanism are given in Figure 4. A critical energy transfer distance R_0 can be defined at which there is an equal probability of energy transfer and spontaneous deactivation of D^*.

$$k_{DA} = k_{FD} \left(\frac{R_0}{R_{DA}} \right)^6 \tag{16}$$

Calculations of R_0 show that it may be >5 nm in favorable cases so that long-range energy transfer is significant when the concentration of acceptor is greater than a critical value $[A]_0$ given by

$$k_{DA}[A]_0[D^*] = k_{FD}[D^*]$$

and the experimental value of R_0 is given by

$$R_0 = \left(\frac{3000}{4\pi N[A]_0} \right)^{1/3} \tag{17}$$

where N is Avogadro's number.

Experimental measurements show that R_0 ranges from 3 to 10 nm. In solution studies k_{DA} may be greater than k_d, the diffusion-controlled rate constant, so that the rate of energy transfer becomes independent of viscosity.

2. Triplet States

In the discussion of energy migration in molecular crystals and along the polymer chain with identical pendant chromophores in Sections III.A and III.B, it was pointed out that triplet energy migration could occur only as a result of electron exchange interactions, in

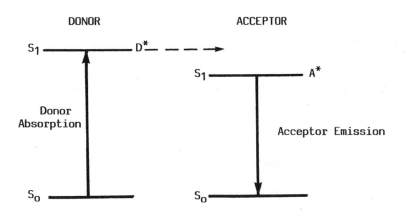

FORSTER   TRANSFER

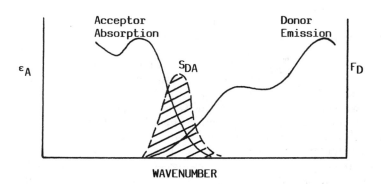

OVERLAP   INTEGRAL

FIGURE 4   Schematic representation of donor–acceptor energy transfer by the Forster mechanism and the overlap integral S_{DA}, showing the energetic and spectral requirements.

the absence of significant spin-orbit coupling. Triplet—triplet
energy transfer between dissimilar immobile molecules would therefore be expected to occur by an analogous mechanism.

$$^3D^* + A \xrightarrow{\quad k'_{DA} \quad} D + {}^3A^*$$ (18)

Since the exchange mechanism requires the overlap of the molecular
orbitals of the excited donor with those of the acceptor, transfer
will ocur over short distances (1.5 to 2 nm).

$$k'_{DA} = S'_{DA} \exp \left(\frac{-2R_{DA}}{L} \right)$$ (19)

where
$\quad R_{DA}$ = separation between D^* and A
$\quad L$ = effective size of the molecules

and S'_{DA} is a spectral overlap integral in which the emission of the
donor and the absorption of the acceptor are both normalized to
unity, so that, unlike Coulombic interaction [Equations (14) and
(15)], the rate constant for triplet energy transfer k'_{DA} does not
depend on the magnitude of the molar absorptivity. Again, a
critical energy transfer radius R_0 can be defined for which k'_{DA} =
k_{DP} where k_{DP} is the rate constant for radiative decay of the donor
triplet state. Experimental data such as the following confirm that
the donor and acceptor molecules must be separated by approximately
a collision distance with R_0 values of 1.5 nm being typical.

An ideal system for demonstrating the occurrence of triplet—
triplet energy transfer is a rigid glass containing benzophenone as
the donor and naphthalene as the acceptor [6]. The electronic
states of the two molecules are given in Table 2, and it can be seen
that because of the large $S_1 - T_1$ energy gap in naphthalene its
excited states bracket those of benzophenone. It is thus possible
to excite the triplet state of benzophenone through its singlet
absorption spectrum without any light being absorbed by naphthalene.
As the concentration of naphthalene in the glass is increased, it is
found that the benzophenone phosphorescence is quenched and
replaced by that of naphthalene. This can only result from direct
triplet—triplet energy transfer as described by Equations (18) and
(19).

This system, suitably modified for the polymer solid state, has
been used to examine energy transfer phenomena in the polymer.
Many real polymer systems, however, do not have well-separated
singlet and triplet states, so that there is often uncertainty in
deciding if singlet or triplet energy transfer is occurring. An
obvious impact of energy migration, trapping, and transfer phenomena on luminescence spectroscopy of solid polymers is that, as

molecular crystals, the emission spectrum of the polymer may be dominated by small amounts of low-energy impurity centers which are populated by energy migration and transfer from the chromophores that are the repeat unit of the polymer.

D. The Polymer Solid State

This chapter is concerned with the general requirements and mechanisms for luminescence from a <u>solid</u> polymer. We have so far considered the photophysical properties of an isolated chromophore and then the situation when it is on a polymer chain where the local chromophore concentration is high. In discussing the requirements for electronic energy migration and transfer, we have assumed that the chromophores will be immobile over the excited state lifetime. In contrast, for small molecules in solution, we have considered the chromophores to be freely diffusing so that Stern-Volmer kinetics apply. The true situation for many polymers at ambient temperatures is between the extremes of a freely diffusing chromophore and one held rigidly in a matrix. A consequence of the viscoelastic nature of polymers is that the properties will be sensitive to temperature.

A solid polymer may be either totally amorphous as in atactic poly(methyl methacrylate) or semicrystalline as in polyethylene, in which both crystalline and amorphous regions occur. The temperature dependence of the stiffness (or Young's modulus E) of an amorphous polymer is shown in Figure 5. At low temperatures, polymers are brittle with a modulus characteristic of many glassy materials. At very high temperatures the material flows and behaves as a viscous liquid. Between these extremes is a rubbery region in which the material has a relatively low modulus and may undergo significant deformation, but because of the entanglement of the polymer chain is mostly recoverable. The temperature at which the transition from glassy to rubbery behavior occurs is known as the glass transition temperature (T_g). A consequence of the long-chain nature of polymers is the difficulty in packing the chains in the solid, so that there is a significant free volume due to this unoccupied space. As temperature is increased, there will be an increase in this free volume and particular motions of the polymer chain will then become possible. At low temperatures, motions such as side group rotation which require a small free volume are observed. As shown in Figure 5, this corresponds to the γ relaxation of the polymer. At higher temperature some backbone mobility is achieved, possibly through a crankshaft type of motion. In many polymers this is the β relaxation. As may be seen from Figure 5, there is only a small change in modulus of the polymer through either the γ or β relaxations, although dynamic mechanical measurements show clear maxima in the loss factor at these temperatures, indicating that an energy-absorbing process is taking place. The activation

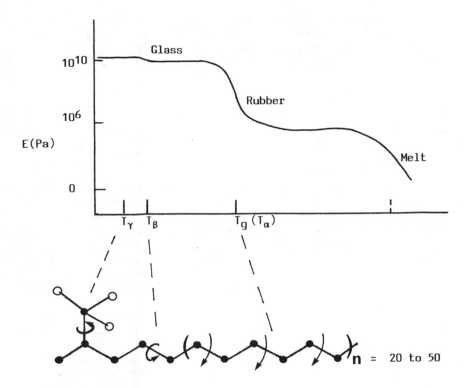

FIGURE 5 Schematic diagram of the temperature dependence of
stiffness (E) of an amorphous polymer and the types of chain motion
that may lead to the γ-, β-, and α-relaxation processes.

energies for these relaxations are generally small, 4 to 20 kJ/mol,
compared to the value of 40 to 120 kJ/mol for the next major relaxation which occurs at the glass—rubber transition temperature (α
relaxation, T_g). The relaxation corresponds to a sharp increase
in the temperature dependence of free volume and results from the
concerted motion of chain segments of length from 20 to 50 carbon
atoms. Translational motion of the chains is still inhibited by
entanglements until the melt or flow temperature where the polymer
behaves as a viscous liquid has been attained.
    It is of obvious importance in the study of polymer photophysics
and luminescence phenomena to know the extent of molecular motion
that is possible at the temperature of investigation. A semicrystalline polymer, solid at room temperature, may have quite high segmental mobility in the amorphous region if it is above T_g. It has
been suggested that, in such a system, a chromophore may have
the same degree of freedom as it would if the polymer were in dilute

solution [12]. Bimolecular processes such as quenching may therefore be of considerable importance. In addition, the consideration
of energy migration and transfer as being either of a freely diffusing chromophore or in an immobile rigid matrix must now be modified
to consider a system of intermediate mobility in which long-range
energy transfer may occur but the centers are also able to diffuse.

In this case Equation (11) is modified to incorporate the relative
molecular diffusion coefficient D of the donor and acceptor (or
quenching molecule and chromophore involved in the migration or
transfer process).

$$\bar{l} = [2(D + D_E)\tau_E]^{1/2} \tag{20}$$

where D_E as before is the diffusion coefficient for excitation energy
migration. Depending on the value of $\bar{l}$, the observed kinetics will
be seen to follow either

1. The Forster Equation (16) for immobile donor and acceptor $\bar{l} \ll R_0$
2. The Stern-Volmer Equation (7) for highly mobile donor and
    acceptor $\bar{l} \gg R_0$

These are the two extremes discussed at the beginning of this
section. The situation where intermediate kinetics apply has been
considered in detail for singlet energy migration and transfer by
North and Treadaway [13]. In studies of poly(N-vinylcarbazole) in
toluene, values of 9×10^{-5} cm^2/sec were obtained for D_E compared
to a value of D of 1×10^{-5} cm^2/sec. In a study of triplet energy
migration in poly(vinylbenzophenone) a value of D_E of 3.3×10^{-5}
cm^2/sec was obtained [14]. This is greater than an earlier value
for D_E of 1.8×10^{-5} cm^2/sec for poly(acrylophenone) under the
same conditions, which is on the order of the molecular diffusion
coefficient D [15].

The sensitivity of the luminescence properties to the mobility
of the polymer means that the relaxation processes of the solid
polymer may be readily studied over a wide temperature range at
a frequency that is given by the rate constant for the photophysical
process. Examples include excimer formation, excited state quenching, and donor—acceptor energy transfer—all of which have been
used to determine relaxations up to T_g. Some of these are outlined
in Section IV.

E. Photochemical Pathways

The discussion of the fate of excitation energy in small molecules,
molecular crystals, and solid polymers has so far considered only
the photophysical pathways. The only processes competing with

luminescence have been radiationless intra- or intermolecular conversion of energy such as intersystem crossing, energy migration, and energy transfer. However, there are many <u>photochemical</u> processes that may also occur. The energy of the excited state molecule may be channeled into bond-breaking reactions or the excited molecule may abstract an atom from an adjacent molecule or from itself to undergo a photochemical reaction. If the initial photon absorbed is sufficiently energetic, then photoionization may occur, leading to charge separation and trapping of the species in the polymer matrix. If photodissociation occurs, then free radicals may be formed which may be trapped or initiate further reactions.

In many cases the photochemical process may be simply regarded as an additional unimolecular or bimolecular radiationless reaction in competition with the photophysical rates. The fundamental properties of quantum yield and lifetime [Equations (2) to (5)] may be simply modified, e.g., the lifetime of the excited state will be reduced to

$$\tau = \frac{1}{k_r + k'} \tag{21}$$

where k' is the sum of the rate constants for the photophysical processes and k_r is the photochemical rate constant. An examination of the photophysical rate constants and quantum yields enables an understanding of the photochemical processes of the polymer as well as some measure of control of the photochemistry by changing the photophysical pathway, for example, by increasing the intersystem crossing rate through enhanced spin-orbit coupling.

The photochemistry of carbonyl-containing polymers has been studied in great detail [12] both because of the wide range of photochemical reactions available from aliphatic and aromatic ketones and also the importance of trace oxidation products on the performance of commercial polymers. Guillet [12] noted that the products of the photochemical reactions of ketones arise from four primary reactions:

1. Alpha cleavage such as the Norrish type I reaction of aliphatic ketones.
2. Hydrogen atom abstraction—either intermolecularly, as in the photoreduction of benzophenone, or intramolecularly, as in the Norrish type II reaction of ketones containing a γ-hydrogen atom. Cyclization may compete with this reaction.
3. Charge transfer complexation.
4. Elimination of an α substituent such as a halogen atom. This will replace the Norrish type I or II reaction.

The Norrish type I reaction of aliphatic ketones occurs either from S_1 ($^1n\pi^*$) or T_1($^3n\pi^*$) excited states. The low quantum yield of fluorescence and phosphorescence ($\phi < 10^{-2}$) and the short triplet lifetime ($\sim 10^{-5}$ sec) are a consequence of the high rate costants for the photochemical reaction k_r [Equations (21) and (22)].

$$R-\overset{\overset{\displaystyle O}{\|}}{C}-R' \xrightarrow{\ k_r\ } R'-\overset{\overset{\displaystyle O}{\|}}{C}\cdot + R\cdot \qquad (22)$$

when R is a methyl group k_r is $\sim 10^3$ sec^{-1} whereas if R is a tertiary butyl group this increases to 2×10^9 sec^{-1}.

If R' is an aromatic group, then there is an increase in the rate constant for intersystem crossing k_{ISC} resulting in a decrease in the rate of photoprocesses from S_1. The value of k_r for reaction (22) is decreased by a factor of 100 as a consequence. The photo-chemistry of aromatic carbonyl compounds occurs primarily from the triplet state T_1. Reactions such as hydrogen atom abstraction are much more efficient if this is a $^3n\pi^*$ rather than a $^3\pi\pi^*$ state. In benzophenone, the rate constant k_r for this bimolecular reaction with cumene is 1.1×10^6 dm^3 mole^{-1} sec^{-1} [14]. If the alkyl group chain length R is at least three carbon atoms and contains a γ-hydrogen atom, then the abstraction occurs intramolecularly—the Norrish type II reaction.

$$ \xrightarrow{\ k_r\ } \qquad + \qquad \qquad (23)$$

This reaction will occur in both aliphatic and aromatic ketone-contain-ing polymers and results in chain scission without the production of free radicals, in contrast to the Norrish type I reaction.

When considering the photochemical reactions in the solid polymer, the temperature dependence of the chain mobility will affect the reaction efficiency. Again, ketone-containing polymers have been the most studied and the following results are typical. Norrish type I reactions often have a greatly reduced quantum yield in both solution and the solid state because of radical recombination and are strongly temperature-dependent [16]. Norrish type II reactions dominate and are less sensitive to temperature above T_g. However, the intramolecular H-abstraction reaction of the Norrish II process requires the formation of a cyclic six-membered ring as a reactive intermediate. There must therefore be sufficient free volume avail-able in the solid state for the formation of this intermediate within

the excited state lifetime.   In a study of the quantum yield of chain
scission in a styrene-phenylvinylketone copolymer [17] it was found
that, below T_g, the yield decreased fourfold.   Reaction below T_g is
thus possible only for those molecules having a preexisting conformation allowing formation of the intermediate.   In contrast to this
result, a study of the photo—Fries rearrangement [12] of aromatic
esters such as poly(phenyl acrylate) has shown a quantum yield of
0.4 in the solid independent of temperature to as low as 21°C,
which is well below T_g (51°C).   The quantum yield decreased below
21°C but was still measurable at T_γ (-73°C).   This result and the
low activation energy suggest that only a very small free volume is
required for reaction, which implies a caged-radical mechanism.

The luminescence properties of the solid polymer may be used
to provide information about the rates and mechanisms of the photochemical reactions occurring from the excited states [12].   Quenching studies have revealed the mechanism of action of photostabilizers
and rationalized data for photodegradation of commercial polymers
[18—20].   In all studies of polymer luminescence, the possible side-
effects of photochemical reaction during the course of the experiment
must be considered.   Berlman [21] noted molecules showing photochemical effects during the measurement of fluorescence spectra of
aromatic compounds in solution, and similar effects may be expected
in polymers.   Photochemically sensitive compounds included brominated
compounds, amines, phenols, quinones, and unsaturated compounds,
many of which will generate free radicals on UV irradiation.   The
subsequent reaction of these free radicals may generate chemiluminescence and thus the photochemistry of the polymer may produce
a new form of luminescence.   This, together with the photoluminescence experiments, is discussed in the next section.

IV.   THE POLYMER LUMINESCENCE EXPERIMENT

Luminescence experiments on solid polymers may be designed to
obtain the fundamental information on molecular excited states as
outlined in Figure 1 and Table 1, or to study solid state processes
such as energy migration and trapping, excited state quenching, as
well as other energy transfer phenomena.   In most cases, only a
limited number of luminescence properties are measured.   These are

* Total intensity of light emitted following an excitation process
* Wavelength dependence of this emission, i.e., the luminescence
    spectrum
* Decay properties of the emission, i.e., the lifetime of the excited
    states
* Polarization properties of the luminescence
* Excitation wavelength dependence of the luminescence, i.e., the
    excitation spectrum

In the solid state these will generally be sensitive to the temperature, atmosphere, and physical state of the polymer. The experimental requirements for obtaining these spectral properties depend on the intensity and lifetime of the emission being studied and the method of exciting the luminescence. In the following sections we will consider the characteristic features of the solid state luminescence process, contrast it with the emission properties from the isolated molecule or solution of the polymer, and relate the experimental results to the photophysical concepts developed in the previous section.

A.   Fluorescence Spectra

The fluorescence spectra of solid polymers, as reported in the literature, have been obtained either with commercial spectrofluori- meters or the equivalent research apparatus constructed with modular components of higher resolution and performance. In most cases the fluorescence is excited with a beam of UV radiation isolated from a high-intensity xenon or mercury vapor discharge lamp by an excita- tion monochromator or suitable filters. In solution studies the fluorescence is gathered at 90° to the exciting light to minimize the problems of scattered radiation. This is not possible with a thin film or polymer chip, and the fluorescence spectrum of a solid poly- mer is routinely obtained with a spectrofluorimeter containing a solid sampling accessory. In many systems the sample is mounted at 45° to the exciting beam but, depending on the nature of the polymer, this may lead to substantial scattering of exciting light. Special accessories are available that allow normal incidence of the exciting light on the solid film and use small mirrors to gather the emitted fluorescence and focus it on the entrance slit of the analyz- ing monochromator. A simple technique that has been widely used is to measure emission from the back face of the sample so that the specularly reflected exciting light is directed away from the analyz- ing monochromator [37,38].

The fluorescence spectrum obtained with a scanning monochro- mator and photomultiplier will be modified by the spectral sensitivity function of the detector as well as the wavelength dependence of the transmission efficiency of the monochromator. While some commercial instruments allow correction for these in either real-time or by computer correction of acquired spectra, much of the reported litera- ture is of uncorrected spectra. While this will not affect the measured spectral band positions, it may affect relative band inten- sities if the instrument is operated in a region where the spectral sensitivity function changes rapidly with wavelength.

A typical solid state fluorescence spectrum from a vinyl aromatic polymer (type B in the nomenclature of Table 3) is shown in Figure 6. This shows the fluorescence from thin films of poly(1-vinyl-

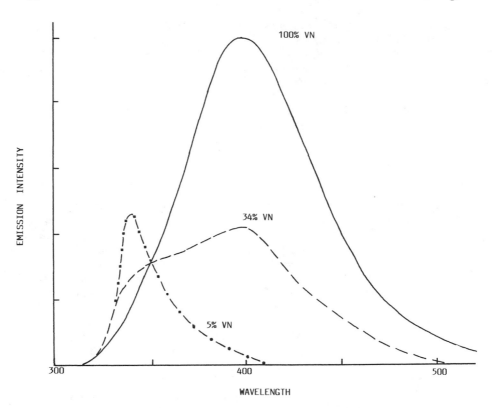

FIGURE 6 Solid state fluorescence spectrum at room temperature of
poly(1-vinylnaphthalene) and poly(1-vinylnaphthalene-co-styrene)
with the percentage vinylnaphthalene as shown. (Adapted from
Ref. 23.)

naphthalene) and copolymers with styrene at room temperature. The
dominant feature is seen to be a broad, structureless emission which
has been attributed to excimer emission [22,23]. The process of
excimer formation was discussed in Section II, and in small molecules
in solution the geometrical requirement is a planar sandwich of the
aromatic groups with a separation of ∿3 Å. This is just the molecular separation in vinyl aromatic polymers.

     In the interpretation of the fluorescence data in Figure 6 at
room temperature and 77K, Fox suggested that both intermolecular
as well as intramolecular excimers could form in the solid state [23].
In addition, the dominance of excimer emission in the fluorescence
could be a consequence of the large number of available excimer-
forming sites or an indication of efficient singlet energy migration.
In order to examine this, David [24] studied the solid state fluores-

cence of copolymers of the structure below in which the number of
carbon atoms separating the phenyl groups was varied from 3 to 6.
No excimer emission was observed.

$$\left(\!-CH_2 - \underset{\underset{\bigcirc}{\overset{CH_3}{|}}}{C} - (CH_2)_m - \underset{\underset{\bigcirc}{\overset{CH_3}{|}}}{C} - CH_2 \!-\!\right)_n$$

This implied that intermolecular excimer formation could not occur.
This was supported by studies of alternating copolymers of styrene
and methyl methacrylate in which no excimer emission was observed
[25]. The extreme sensitivity of excimer emission to the conformation
of the polymer chain has been exploited in studies of polyblend
compatibility. Any chain interpenetration occurring in a miscible
blend, one component of which is an excimer-forming polymer, will
result in a decrease in the number of available excimer-forming sites
and thus a decrease in excimer fluorescence intensity. Phase
separation thus gives higher excimer fluorescence. Gashgari and
Frank exploited this in a series of studies of polymer compatibility,
and in the particular example of poly(2-vinylnaphthalene) dispersed
in a series of methacrylates, excimer fluorescence was weakest where
the solubility parameters were closest [26]. Recent studies have
suggested that the increase in excimer emission may result from
contraction of the polymer coils [27]. Guillet [12] pointed out that
the use of the change in the ratio of excimer emission to monomer
emission to indicate chain interpenetration may be further complicated
by the appearance of nonnearest neighbor excimers from chromophores
on the same chain but separated by a coil or loop.

In the use of solid state fluorescence to measure polyblend
compatibility, the complications in the interpretation of the excimer
emission from type B polymers in terms of chain interpenetration
may be avoided by using fluorescent labels on a nonfluorescent
polymer chain (type A, Table 3). If the labels are chosen to form
a donor—acceptor pair [Equation (13)], then if the two molecules on
different polymer chains are a distance R_{DA} apart, energy transfer
will occur at a rate given by Equation (14). Morawetz [28] put
these ideas into practice in the study of poly(methyl methacrylate)-
poly(styrene-co-acrylonitrile) blends and some of the results are
presented in Chapter 2, Figure 16.

The study of energy migration and transfer by fluorescence
spectroscopy has not been carried out to the same extent in the
solid polymer as it has in solution because of the difficulty in separating intermolecular and intramolecular perturbations. David and

Baeyens-Volant [22] reviewed the literature on solid state fluorescence and concluded that the present knowledge of singlet energy
migration and transfer processes is only fragmentary and has come
from only steady-state intensity measurements. In spite of this,
some important applications, as discussed above, have been achieved.
It is considered that further development will require time-resolved
fluorescence spectroscopy, as has been applied to polymer solutions
[5,29].

B. Phosphorescence Spectra

Under the normal conditions of observation of luminescence spectra
of solid films in air, phosphorescence emission is either very weak
or totally absent. As discussed in Section II.B, this results from
bimolecular quenching of the long-lived triplet state, in particular
by molecular oxygen. The study of the phosphorescence spectra
and lifetimes of vinyl aromatic polymer films has therefore been
performed at low temperature in vacuum to minimize bimolecular
processes (including photochemical reaction) and thus obtain the
unimolecular rate constants [11]. The phosphorescence lifetime in
the absence of quenching may range from milliseconds for carbonyl
chromophores to several seconds for aromatics, so that the emission
from the triplet state may be separated from fluorescence and
scattered radiation by a suitable time delay before the spectrum is
recorded. This may be achieved by either a rotating shutter or
flash excitation with a gating period before the spectrum is recorded.
Phosphorescence, as shown in Figure 1, is observed at lower energy
than fluorescence.

   When the complete delayed emission spectrum from a solid polymer
is examined, it is found that delayed fluorescence resulting from
triplet—triplet annihilation (Section II.B) is also observed. Webber
[11] noted that both phosphorescence and delayed fluorescence from
poly(2-vinylnaphthalene) are broadened and shifted compared to the
polymer in solid solution and may be assigned to triplet and singlet
excimer emission, respectively. This is shown in Figure 7, in which
the effect of molecular weight on the spectrum is also shown [30].
This was initially interpreted as evidence for triplet—triplet annihilation occurring only between excited states on the same chain (i.e.,
triplet energy migration has occurred but there is little interchain
transfer, even in the solid state). The data could also indicate that
triplet state trapping occurs at chain ends or defects resulting from
chain ends, so that delayed fluorescence increases with molecular
weight due to the increased mean free path for migration of the
triplet excitation [11]. Evidence has been produced for immobiliza-
tion of triplet excitation at low temperatures in poly(1-vinylnaph-
thalene) so that phosphorescence is favored over delayed fluorescence
[31]. Once again emission was attributed to the triplet excimer.

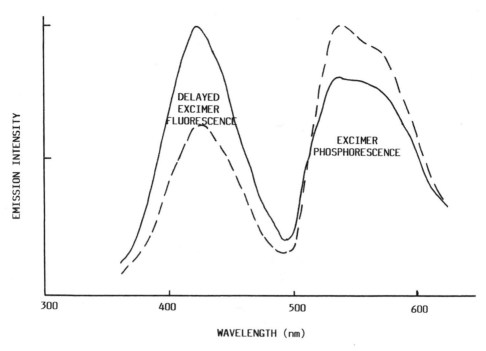

FIGURE 7   Delayed emission spectrum from films of poly(2-vinyl-
naphthalene) showing delayed excimer fluorescence (from 350 to 500
nm) and phosphorescence (from 500 to 630 nm).   The dotted line
represents the result for lower molecular weight polymer. (Adapted
from Ref. 11.)

In contrast to the vinyl aromatics, films of vinyl aromatic ketone
polymers show phosphorescence spectra with clear vibrational pro-
gressions at all temperatures that are very similar to those of the
repeat units [32—34].   Evidence for triplet energy migration in these
polymers, e.g., poly(vinylbenzophenone) or poly(phenylvinylketone)
has come from the measurement of the critical energy transfer radius
R_0 from quenching experiments with naphthalene and the use of
Equation (20), when it is found that the value of R_0 is greater than
would be expected for static quenching [Equation (19)].

In studies of the phosphorescence from commercial polystyrene
films and those prepared in the laboratory under controlled partial
pressures of oxygen [35,36], a structured emission characteristic
of aromatic ketone polymers built on an underlying excimerlike band
was observed (Figure 8).   From lifetime studies and spectral analysis,
the two spectra could be separated and the structured emission, (i),
was assigned by band analysis and comparison with model compounds

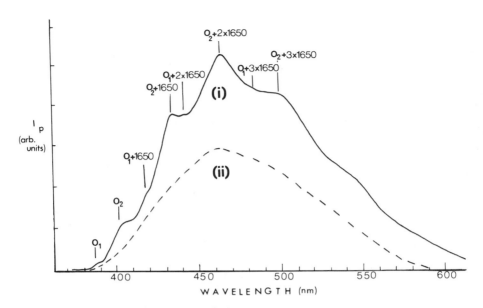

FIGURE 8 Phosphorescence spectrum of polystyrene containing a
low concentration of phenylalkylketone end groups due to incorporation of oxygen during synthesis. Spectrum (i): delay time 1 msec;
showing vibrational progression of carbonyl group. Spectrum (ii):
delay time 75 msec; showing triplet excimer emission. (From Ref.
35.)

to phosphorescence from phenylalkylketone end groups formed by
the incorporation of oxygen in the polymer during polymerization and
processing [35,36]. The broad band, (ii), was assigned to triplet
excimer emission on the basis of the band shift of 6600 cm^{-1} from
the ethyl benzene "free-molecule" spectral origin (in close agreement
with calculations for the shift for benzenelike excimers) and the
lifetime of 1.5 sec, which was too long to be that of a ketone
impurity. This result for polystyrene, in which the intrinsic phosphorescence of the polymer film was masked by impurity emission, is
a direct consequence of the efficient energy migration, transfer, and
trapping of the long-lived triplet state. In many other polymers,
the phosphorescence spectrum of the solid is also not characteristic
of the repeat unit, but rather results from emission from low concentrations of oxidation products [37]. Further results in this area
are detailed in Chapter 2.

    In addition to the use of phosphorescence spectroscopy to detect
the presence of oxidation products which may be of importance in
the environmental performance of the polymer, it is possible under

favorable circumstances to determine their concentration.   While
quantitative fluorescence spectroscopy is difficult in solid films
because of scattering when the sample is placed at the usual angle
of 45° to the beam of exciting radiation, the study of the delayed
emission allows other geometries to be used.   As the exciting light
is separated from the emission by a time delay, phosphorescence (or
delayed fluorescence) may be observed collinear with the exciting
light so that scattering is no longer a problem and the amount of
emission may be directly related to the amount of light absorbed.

This arrangement is shown in Figure 9, and for low absorbance
by the chromophores the phosphorescence intensity I_p from a film
of thickness d is given by:

$$I_p = k\phi_p(I_0 - I)$$

which at low absorbance

$$\approx k\phi_p I_0 \alpha d \qquad\qquad\qquad (24)$$

where

α = absorption coefficient
ϕ_p = phosphorescence quantum yield
k = fraction of emitted light detected by the apparatus

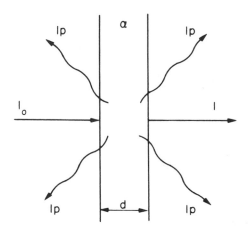

FIGURE 9   Sample arrangement for quantitative phosphorescence
spectroscopy. I_p is measured from the back face of the film.   (From
Ref. 39.)

In terms of a concentration c of chromophores of molar absorptivity
ε

$$I_p \simeq 2.303 \varepsilon c d k \phi_p \qquad \qquad (25)$$

This relation was tested for polystyrene containing known concentrations of 1,3-diphenyl propan-1-one (a model compound for the
carbonyl impurity in polystyrene that was seen in Figure 8). As
shown in Figure 10, a linear relation is obtained and this has been
used to measure both the initial impurity group concentration in
commercial polystyrene ($\sim 10^{-3}$ mol/dm^3) and the oxidation product
concentration during photooxidation [39]. A typical sensitivity for
quantitative analysis of aromatic carbonyl chromophores by phosphorescence is 2×10^{-5} mol/mol compared with typical analytical
limits in polymer spectroscopy of 3×10^{-4} mol/mol for IR; 5×10^{-4}
mol/mol for UV and 9×10^{-4} mol/mol for ^{13}C NMR spectroscopy.
While the major experimental difficulty in quantitative phosphorescence spectroscopy is the requirement for low temperatures and
either vacuum or an inert atmosphere, in some polymers room

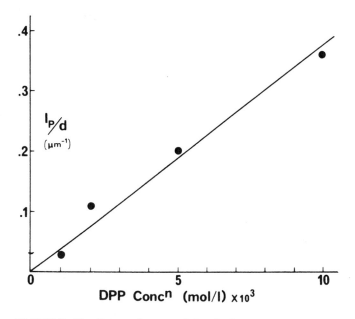

FIGURE 10 Dependence of back-face phosphorescence intensity from
a polystyrene film on the concentration of added 1,3-diphenyl propan-
1-one (DPP). (From Ref. 39.)

temperature phosphorescence in air can be observed. This is believed to arise from the low solubility and diffusion coefficient of oxygen in some synthetic polymers such as the polyamides and natural polypeptides such as wool—both of which phosphoresce at room temperature in air. For example, in nylon 6,6 fibers, the diffusion coefficient of oxygen is 1.25×10^{-11} cm^2/sec leading to a diffusion-controlled rate constant of 9.4×10^{-4} dm^3 mol^{-1} sec^{-1} in the Stern-Volmer Equation (7). Since the solubility of oxygen is 1.4×10^{-3} mol/dm^3 at a partial pressure of 0.2 atmosphere, and the triplet lifetime is ~ 75 msec, the quantum yield of phosphorescence from the bulk polymer is reduced only by a factor of 11 when measured in air compared to vacuum. This greatly simplifies the experimental technique and has been applied to the degradation of nylon fibers [19], although the possibility of a photochemical pathway being introduced when the polymer is irradiated in air at room temperature must be considered. Such a situation is discussed later in Section IV.E.

C. Excitation Spectra

The excitation spectrum of either fluorescence or phosphorescence is obtained if the variation in emission intensity from the polymer at a fixed wavelength (or over a broad emission band isolated by a filter) is measured as the excitation wavelength is changed. In most spectrofluorimeters this involves scanning the excitation monochromator over the wavelength range of the absorption spectrum of the polymer and measuring the resultant change in emission intensity. As can be seen from the state diagram of Figure 1, because of rapid internal conversion (or intersystem crossing in the case of phosphorescence) the emission spectrum will be independent of excitation wavelength. The fluorescence intensity $I_F(\lambda)$ emitted for each excitation wavelength λ is given from Equation (2) by

$$I_F(\lambda) = \phi_F I_a(\lambda)$$

$$= \phi_F I_0(\lambda) \, (1 - 10^{-\varepsilon cd}) \tag{26}$$

if $\varepsilon cd \ll 1 \simeq 2.303 \phi_F I_0(\lambda) \varepsilon cd$

Thus if $I_F(\lambda)/I_0(\lambda)$ is plotted as a function of wavelength, it should generate an absorption spectrum of the emitting chromophore which is the mirror image of the emission spectrum. The raw excitation spectrum consists only of $I_F(\lambda)$ scanned over the wavelength range of interest. Since most spectrofluorimeters are single-beam devices, I_0 varies substantially with wavelength due to the emissivity of the source and transmission function of the excitation monochromator. Uncorrected excitation spectra will therefore show a different intensity distribution from the absorption spectrum, although peaks will occur at the same wavelength. The correction function $I_0(\lambda)$ may be most readily measured by replacing the sample with a "quantum counter." This is a concentrated solution of a fluorescent dye (e.g., rhodamine B) or a solid crystal (e.g., sodium salicylate) which has a constant quantum yield and absorbs all incident quanta over a wide band. The correction curve may then be stored and spectra corrected routinely by computer. Alternatively, some instruments are able to correct in real time by sampling part of the excitation beam continuously with a separate quantum counter and photomultiplier, and correcting the excitation spectrum with a feed-back circuit [29].

An advantage of excitation spectra over absorption spectra is the high sensitivity of the technique. For example, $T_1 \leftarrow S_0$ spectra of aromatic hydrocarbons and molecular crystals have been readily detected by phosphorescence excitation spectroscopy of thin samples (100 μm) whereas direct absorption spectra for the spin-forbidden transition would require a pathlength of many centimeters because of the low molar absorptivity ($\varepsilon \sim 10^{-4}$ dm^3 mol^{-1} cm^{-1}). The $S_1 \leftarrow S_0$ excitation spectrum of both fluorescence and phosphorescence has been of great value in determining the emitting species in type A polymers [37]. However, it must always be noted that because of efficient energy transfer in the solid state, the excitation spectrum may be that of the most strongly absorbing species and is not necessarily that of the emitting chromophore. This has been exploited in aromatic crystals by incorporating a high quantum efficiency luminescent dopant in an otherwise nonluminescent solid and then measuring the absorption spectrum of the host from the excitation spectrum of the dopant luminescence [1].

D. Polarization Properties and Time-Resolved Spectra

The absorption and emission spectra of aromatic molecules such as naphthalene are polarized along particular molecular axes, depending on the nature of the electronic transition [1]. For example, the $S_1 \leftrightarrow S_0$ ($\pi\pi^*$) transitions of naphthalene correspond to transition moment vectors lying in the plane of the molecule while $T_1 \leftrightarrow S_0$ transitions are polarized out of plane. In contrast, for the $\pi\pi^*$-excited states of the carbonyl group, the $S_1 \leftrightarrow S_0$ transitions are out-of-plane and the $T_1 \leftrightarrow S_0$ transitions are in-plane polarized. Consequently, if the luminescence is excited with plane-polarized light and the polarization properties of the emission are measured, the degree of polarization P is given by

$$P = \frac{I_\parallel - I_\perp}{I_\parallel + I_\perp}$$

and the emission anisotropy r is given by

$$r = \frac{I_\parallel - I_\perp}{I_\parallel + 2I_\perp}$$

where $I_\parallel$ and $I_\perp$ are the luminescence intensities when passed through an analyzing polarizer that is oriented parallel and perpendicular, respectively, to the excitation polarizer. Experimentally this is achieved by placing a fixed polarizer before the sample (e.g., Polaroid sheet HNP'B) and a rotatable polarizer before the analyzing monochromator.

For randomly oriented molecules that do not rotate within the excited state lifetime, the value of P is determined by the angle between the transition moment vectors for absorption and emission. If this is 90°, P has the value of -0.33 and if it is 0°, P will be +0.5. If the rotational relaxation time τ_r of the chromophore on the polymer chain is comparable to the lifetime τ_F of the excited state, then partial relaxation occurs and analysis of the polarization properties can yield information concerning this rotational motion [2,40]. Studies of polymers labeled with isolated fluorescent groups have yielded measurements of the segmental motion of the polymer chain in solution by measuring the fluorescence anisotropy r on a nanosecond time scale [29,40]. In this way relaxation times of the chain ends and central segments may be compared. Of particular interest has been the conformation of polyelectrolytes [41]. Time-resolved fluorescence spectra have provided further information on the pathways of energy migration and trapping, particularly excimer formation [4]. The rate constants for excimer formation and dissociation in solution have been obtained in this way and used to test current theories [3,4].

Of more relevance to studies of solid state properties are polarization measurements of bulk polymers. These have been aimed at either measuring relaxation processes in the amorphous solid state [22,42] or determining the extent of orientation of the polymer chain [43] so complementing x-ray diffraction and birefringence measurements. The relaxation processes that can be probed by polarization anisotropy will depend on the lifetime of the excited state. Because of the short fluorescence lifetime, motions of frequency 10^8 to 10^{10} Hz may be studied whereas phosphorescence polarization will probe relaxations in the range 10^{-1} to 100 Hz. In a recent study [42], the temperature dependence of P and thus τ_r of copolymerized naphthalene and acenaphthylene enabled the α and β relaxations of n-butyl acrylate to be determined. Polarization studies of amorphous polymer films have also been used to test the theories of energy migration and trapping in polymers since polarization properties will be lost if excitation energy migrates during the excited state lifetime

[7]. For example, when polarization measurements were carried out [35] on the time-resolved phosphorescence spectra of polystyrene films as shown in Figure 8, it was found that the structured emission (i) showed a positive value of P consistent with emission from a $^3n\pi^*$ state of an aromatic carbonyl group but the triplet excimer emission (ii) was depolarized. This implies triplet energy migration followed by trapping at an excimer-forming site.

The orientation and deformation of amorphous and semicrystalline polymers has been studied by polarized fluorescence of a probe molecule either copolymerized or merely trapped in the polymer matrix [43]. By using a probe molecule with a high aspect ratio, the molecule adopts a preferred orientation with respect to the chain direction. The measured angular distribution of polarized fluorescence may be compared with different theoretical values for different extents of orientation to determine the overall chain orientation in the amorphous region. In one particular study [44] an anthracene derivative was copolymerized in the middle of the polymer chain such that the transition moment vector for fluorescence was in the chain direction. The changes in polarization during deformation and swelling of the network were then studied.

E. Chemiluminescence

In Section III.E it was noted that the process of absorption of UV radiation may lead to photodissociation in the excited state and the formation of free radicals. If this process takes place in the solid polymer in the presence of oxygen, a very weak luminescence may be detected that persists for minutes or even hours after irradiation has ceased. Since photoluminescence decays over the time scale of seconds, such long-lived emission cannot reflect the lifetime of the singlet or triplet state but must result from the decay of the free radicals generated by the absorption of radiation. This photochemically generated luminescence is just one example of the chemiluminescence observed during the oxidation of polymers [45]. One of the most closely studied processes is the chemiluminescence arising from thermal oxidation at temperatures above 100°C. Because of the low quantum efficiency ($\phi \sim 10^{-9}$) of chemiluminescence compared to that of fluorescence and phosphorescence ($\phi \sim 1$ to 10^{-4}), it is normally not observed in routine measurements of photoluminescence spectra. The recent development of stable photon-counting techniques for measuring low light levels has made reliable chemiluminescence measurements a simple process, and commercial apparatus for polymer studies is now available. Several models of photon-measuring chemiluminescence apparatus are available from the Tohoku Electronic Industrial Company, Sendia, Japan.

In the framework of the electronic state diagram (Figure 1), chemiluminescence involves the usual photophysical processes, but

the excitation mechanism involves a chemical reaction.   The excess
energy of the reaction, instead of being lost as heat to the surround-
ings, appears as electronic excitation of the molecules to either S_1
or T_1 [46].   Depending on the normal photophysical rate parameters,
fluorescence or phosphorescence will be observed.   The study of
chemiluminescence from small organic molecules in solution has shown
that the more efficient reactions involve either electron transfer (in
the absence of oxygen); decomposition of peroxides or a reactive
dioxetane intermediate; or the formation and emission of singlet
oxygen [6].   As seen from the transition state diagrams in Figure
11, in principle both endothermic and exothermic reactions may give
rise to chemiluminescence but in the former case the energy of the
quantum $h\nu$ will be too small for the emission of visible radiation
since $h\nu \leq \Delta H^{\neq} - \Delta H$.   All of the chemiluminescent reactions are
exothermic and in some cases the quantum efficiency for excitation
lies between 0.1 and 1.

   If a fluorescent molecule is present with a lower lying electronic
energy level than the excited state produced in the chemical reaction,
then donor—acceptor energy transfer (Section III.C) may occur with
the result that the chemiluminescence spectrum is now that of the
foreign fluorescent molecule.   This has been used to enhance the
yield of chemiluminescence [46].   Conversely, excited state quenching phenomena may decrease the yield of chemiluminescence and if
the emitting state is a triplet, the quantum yield may be only 10^{-6}
of that expected.   Quenching by molecular oxygen is believed the

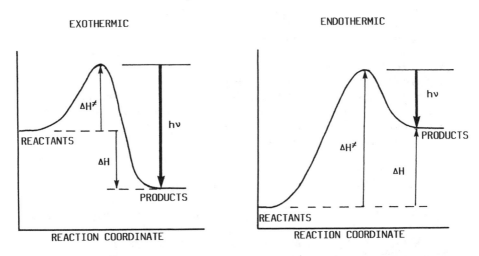

FIGURE 11   Transition state diagrams for potential chemiluminescent
reactions.   Only the exothermic reaction leads to chemiluminescence
in the visible region.

main reason for the low yield of chemiluminescence ($\phi \sim 10^{-8}$ to 10^{-9}) in the solution oxidation of hydrocarbons [47]. A similarly low yield is observed in the oxidation of polymers and in most studies the mechanisms established in solution have been successfully applied. Measurement of even low-resolution chemiluminescence spectra are difficult because of the low light intensities involved, but spectral distributions from most polymers show close similarity to the $T_1 \rightarrow S_0$ emission from carbonyl oxidation products. These may arise in either the initiation, propagation, or termination reaction of the oxidation which is believed, in most common polymers, to be initiated by thermolysis or photolysis of trace amounts of hydroperoxides. These mechanisms, as well as kinetic schemes describing the time evolution of chemiluminescence intensity throughout the oxidation of the polymer under steady-state and nonstationary conditions, are described in the several chapters in this book that explore the mechansims and applications of chemiluminescence.

The particular advantage of chemiluminescence is that it provides a sensitive technique for monitoring free-radical populations in solid polymers. The effect of the environmental parameters of temperature, atmosphere, radiation, and stress may then be assessed. The information generated is thus complementary to studies of radical species by electron spin resonance and chemical spin-trapping techniques [48,49].

F.   Thermoluminescence

Thermoluminescence (or radiothermoluminescence) is the weak light emitted during the heating of a polymer sample that has previously been subjected to ionizing radiation at low temperature. It is distinguished from photoluminescence phenomena by the observation that it may occur after leaving the irradiated sample for times ranging up to days prior to heating. In contrast to chemiluminescence, the maximum thermoluminescence intensity generally occurs at temperatures below 250K [50]. The apparatus required to study thermoluminescence is similar to a chemiluminescence apparatus except that provision must be made for high-energy irradiation of the sample while held at low temperature [51]. Emission may be observed by programming the temperature from 77 to 300K and measuring the resultant glow curve or by monitoring the decay in intensity when held at a particular temperature in this range.

The general mechanism for the production of thermoluminescence is believed to be the migration and recombination of trapped species generated by the ionizing radiation. Spectral analysis shows that emission covers the range 300 to 600 nm and in the polyolefins is attributed to fluorescence and phosphorescence from a carbonyl group, although in several studies the presence of aromatic impurities as emission centers was indicated [52]. Considerable research has been carried out regarding the process of thermally assisted electron escape and migration prior to recombination and emission of light from the impurity center [50]. Three mechanisms have been considered:

1. Thermal excitation out of stable traps. This model envisages the electron trap site as either a cavity created by a chain defect or an impurity with a positive electron affinity such as oxygen. The measured activation energies for emission will then correspond to the depths of the traps.
2. Trap break-up by chain motion. Progressive heating of the polymer through the γ, β, and α transitions (Section II.D) would be expected to break up electron traps such as chain defects and facilitate electron release from impurity molecules such as oxygen. The activation energies for emission will then correspond to the particular chain motion commencing at each transition temperature and not to the depth of the potential well of the electron trap. However, Fleming and Hagekyriakou [50] noted that correlating these thermoluminescence data with methods for measuring loss peaks in the polymer such as dielectric response, which operates at 1 kHz or higher, may not be feasible because the equivalent operating frequency of the thermoluminescence measurement is of the order of 10^{-3} Hz.
3. Electron tunneling from traps. The recombination is viewed as a resonant process in which the electron does not have to be thermally detrapped or released by increased molecular motion. This has been used to explain isothermal decay data.

In the study of the solid state properties of polymers by thermoluminescence, the use of model 2 above is of the most value. While the correlation of major glow peaks with temperature transitions is certain, there are some complicating factors and further research into the mechanism of the phenomenon is continuing [55]. A separate chapter of this book is devoted to a detailed discussion of polymer thermoluminescence.

G. Luminescence from Small-Molecule Probes

In the previous sections the luminescence properties of polymers in which the chromopores form the repeat unit (type B) or are present as isolated groups ether deliberately copolymerized or as adventitious impurities on the polymer chain (type A) have been considered. Particular emphasis has been placed on those experiments in which luminescence may provide information about the polymer structure, free volume, and molecular interactions. If the polymer is not intrinsically luminescent, similar information can be obtained by incorporating free fluorescent and phosphorescent molecules with photophysical properties chosen to probe particular relaxation,

structural, or deformation processes in the solid state. The probe
molecule may be incorporated at synthesis, when processing, by
casting from a solvent or simply by soaking the polymer in a swelling solvent containing the probe. There may be particular advantages in this approach, such as when the probe may be tailor-made
to study specific interactions. For example, the fluorescence
properties at 5-dimethylamino-1-naphthalenesulfonate have been
studied in cationic copolymers of styrene and vinylbenzene trialkyl
ammonium halides and found to be sensitive to polar interactions in
the polymer [54].

All of the luminescence techniques described in the previous
sections may be applied, and by choice of appropriate excitation
wavelengths, temperature, and atmosphere, it is possible to observe
energy transfer, excimer formation, quenching phenomena, and other
bimolecular processes. Table 4 summarizes just some of the changes
to luminescence properties that may be exploited in the study of
particular relaxation phenomena. Representative probe molecules
are shown below:

1   2   3   4

TABLE 4   Use of Probe Molecules to Monitor Relaxation Processes
in Polymers

Probe	Polymer	Relaxation	Mechanism
1	Polyisoprene Polybutadiene SBR	T_g	Excimer formation increases with free volume [55,56].
2	PMMA PVC-P(vinyl acetate) PS Epoxy resins	T_g	Radiationless rate constant increases with free volume [58,59].
3	PMMA PMA	T_β T_γ T_g	Triplet lifetime decreases due to ester quenching [61].
4	PMMA PS PVA	T_β T_γ	Radiationless rate constant increases with side group motion [62].

Excimer-forming probes have been used to study free-volume changes with temperature [55,56]. Both the intensity and lifetime of excimer fluorescence of the probe molecule 1 (10,10'-diphenylbis-9-anthryl-methyl oxide) are sensitive to the free-volume change at T_g. This concept has been extended by Wang et al. [57] to the study of the viscosity changes occurring in epoxy resins during cure by using 1,3-bis(1-pyrene)propane as the excimer-forming probe. In another approach by the same authors, two fluorescent dyes have been incorporated in an epoxy resin to monitor the crosslinking reaction. One dye is a diphenyl hexatriene derivative (DPH) that shows viscosity-dependent fluorescence; the other, diphenyl anthracene (DPA) is unaffected by viscosity and is used as an internal standard. Typical results with these probes [57] are shown in Figure 12 for the curing of a bisphenol A epoxy resin at 60°C. The internal standard allows for local variations in refractive index and light scattering that alter the absolute emission intensity from the probe.

Perhaps the most detailed studies of the use of fluorescent probes to study the changes in polymer viscosity and free volume during polymerization have been those of Loutfy, who recently reviewed his research [58]. He has used a number of substituted malononitriles such as molecule 2 (julolidene malononitrile), which have been termed "molecular rotors." This and similar molecules based on the malono-nitrile structure have a first singlet excited state with substantial charge-transfer character that undergoes radiationless decay at a rate dependent on the free volume. A sharp decrease in intensity

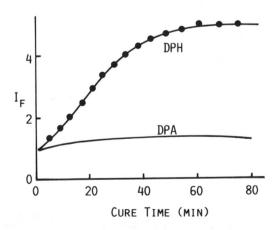

FIGURE 12 Changes in the fluorescence intensity of the probe molecule DPH and the internal standard DPA during the cure of a bisphenol A epoxy resin at 60°C. (Adapted from Ref. 57).

occurs at T_g. Conversely, in a polymerizing system the intensity
increases as viscosity increases up to gelation. Some epoxy resin
systems have been shown to undergo self-probe fluorescence. In
this case the repeat unit contains a structure similar to the molecular
rotors and the intrinsic fluorescence provides a monitor of the viscosity changes during gelation [59]. An interesting application of
the use of a fluorescent probe to follow the viscosity change of a
photopolymer during curing involves the measurement of the polarization anisotropy of a naphthalimide derivative. As discussed in
Section IV.D for labeled polymers, the depolarization is a measure of
the rotational freedom and this will change during gelation [60].

The probe molecules 3 and 4 show triplet state emission properties that are sensitive to relaxation processes of the host polymer.
Benzophenone (Structure 3) shows a change in the triplet decay
kinetics at the β and α relaxation temperatures which was attributed
to quenching by side chain ester groups [61]. The coumaric acid
derivative (Structure 4) similarly shows an increase in the radiationless rate constant with the onset of side group motion, so leading to
a decrease in emission intensity at a characteristic temperature [62].
These applications have built on the original work of Somersall et al.
[63] in which the quenching of phosphorescence was attributed to
the increased accessibility of trace molecular oxygen to the emitting
chromophore as the polymer becomes more mobile. More recently it
was pointed out [12] that the apparent relaxation temperature depends on the partial pressure of oxygen above the film and a more
detailed treatment is required. Conversely, the quenching of
phosphorescence may be used to measure the diffusion coefficient of
the quencher (usually oxygen) in the solid film.

V. CONCLUSION

Fundamental concepts from the photophysics of isolated organic
molecules and organic molecular crystals may be applied with suitable
modifications to describe the luminescence from solid organic polymers.
It is found that solid state fluorescence and phosphorescence spectra
are dominated by excimer emission and impurity emission—both a
consequence of energy migration and transfer in the polymer. The
theoretical description of energy migration, trapping, and excimer
formation in the solid is far from complete but there are many
successful applications of luminescence spectroscopy to practical
polymer problems. All of these take advantage of the great sensitivity of luminescence over absorption spectrophotometry and include
studies of degradation and stabilization, polymer relaxations and
mobility, morphology, and orientation as well as the chemical identification of the polymer and its impurities. The use of probe molecules
increases the flexibility of the techniques and enables specific inter-

actions in the solid polymer to be studied. In addition to excitation by absorption of a photon, the use of ionizing radiation and the energy of chemical reactions are enabling studies of great sensitivity and specificity through thermoluminescence and chemiluminescence of solid polymers. *Graphs, spectra.*

ACKNOWLEDGMENTS

The financial assistance of the Australian Research Grants Scheme and the Defence Science and Technology Organisation for studies of polymer luminescence is gratefully acknowledged.

REFERENCES

1. J. B. Birks, Photophysics of Aromatic Molecules, Wiley-Interscience, New York (1970).

2. D. Phillips, Polymer Photophysics D. Phillips, ed.), Chapman and Hall, London, p. 1 (1985).

3. D. A. Holden and J. E. Guillet, Developments in Polymer Photochemistry Vol. 1 (N. S. Allen, ed.), Applied Science, London, p. 27 (1980).

4. A. J. Roberts and I. Soutar, Polymer Photophysics (D. Phillips, ed.), Chapman and Hall, London, p. 221 (1985).

5. K. P. Ghiggino, A. J. Roberts, and D. Phillips, Adv. Polym. Sci., 40: 71 (1981).

6. J. A. Barltrop and J. D. Coyle, Principles of Photochemistry, John Wiley and Sons, London (1978).

7. S. N. Semerak and C. W. Frank, Adv. Polym. Sci., 51: 31 (1984).

8. D. P. Craig and S. H. Walmsley, Excitons in Molecular Crystals, Benjamin, New York (1968).

9. E. A. Silinish, Organic Molecular Crystals: Their Electronic States, Springer-Verlag, Heidelberg, p. 223 (1980).

10. S. E. Webber, New Trends in the Photochemistry of Polymers (N. S. Allen and J. F. Rabek, eds.), Elsevier, London, p. 16 (1985).

11. S. E. Webber, Polymer Photophysics (D. Phillips, ed.), Chapman and Hall, London, p. 42 (1985).

12. J. E. Guillet, Polymer Photophysics and Photochemistry, Cambridge University Press, Cambridge (1984).

13. A. M. North and M. F. Treadaway, Eur. Poly. J., 9: 609 (1973).

14. E. H. Urruti and T. Kilp, Macoromolecules, 17: 50 (1984).

15. T. Kilp and J. E. Guillet, Macromolecules, 14: 1680 (1981).

16. G. H. Hartley and J. E. Guillet, Macromolecules, 1: 165 (1968)
    F. J. Golemba and J. E. Guillet, Macromolecules, 5: 63 (1972).

17. E. Dan and J. E. Guillet, Macromolecules, 6: 230 (1973).

18. E. D. Owen, Developments in Polymer Photochemistry, Vol. 1
    (N. S. Allen, ed.), Applied Science, London, p. 1 (1980).

19. G. A. George, Pure Appl. Chem., 57: 945 (1985).

20. D. A. Holden, K. Jordan, and A. Safarzadeh-Amiri, Macro-
    molecules, 19: 896 (1986).

21. I. B. Berlman, Handbook of Fluorescence Sprctra of Aromatic
    Molecules, Academic Press, New York, p. 18 (1965).

22. C. David and D. Baeyens-Volant, Polymer Photophysics (D.
    Phillips, ed.), Chapman and Hall, London, p. 115 (1985).

23. R. B. Fox, T. R. Price, R. F. Cozzens, and J. R. McDonald,
    J. Chem. Phys., 57: 534 (1972).

24. C. David, M. Lempereur, and G. Geuskens, Eur. Polym. J.,
    9: 1315 (1973).

25. R. B. Fox, T. R. Price, R. F. Cozzens, and W. E. Echols,
    Macromolecules, 7: 937 (1974).

26. M. A. Gashgari and C. W. Frank, Macromolecules, 14: 1558
    (1981).

27. J. M. Torkelson, M. Tirrell, and C. W. Frank, Macromolecules,
    17: 1505 (1984).

28. H. Morawetz, Polym. Eng. Sci., 23: 689 (1983).

29. L. L. Chapoy, D. B. Du Pre, and D. Biddle, Developments in
    Polymer Characterization, Vol. 5 (J. V. Dawkins, ed.), Elsevier,
    London, p. 223 (1986).

30. N. Kim and S. E. Webber, Macromolecules, 13: 1233 (1980).

31. R. D. Burkhart, R. G. Aviles, and K. Magrini, Macromolecules,
    14: 91 (1981).

32. C. David, W. Demarteau, and G. Geuskens, Eur. Polym. J.,
    6: 537 (1970).

33. C. David, W. Demarteau, and G. Geuskens, Eur. Polym. J.,
    6: 1397 (1970).

34. C. David, W. Demarteau, and G. Geuskens, Eur. Polym. J.,
    6: 1405 (1970).

35. G. A. George, J. Appl. Polym. Sci., 18: 419 (1974).

36. W. Klopffer, Eur. Polym. J., 11: 203 (1975).

37. N. S. Allen, Analysis of Polymer Systems (L. S. Bark and N. S. Allen, eds.), Applied Science, London, p. 79 (1982).

38. C. W. Frank and L. A. Harrah, J. Chem. Phys., 61: 1526 (1974).

39. G. A. George and D. K. C. Hodgeman, Eur. Polym. J., 13: 63 (1977).

40. L. Monnerie, Polymer Photophysics (D. Phillips, ed.), Chapman and Hall, London, p. 279 (1985).

41. K. P. Ghiggino and K. L. Tan, Polymer Photophysics (D. Phillips, ed.), Chapman and Hall, London, p. 341 (1985).

42. I. Soutar and J. Toynbee, Polym. Preprints, 27(2): 338 (1986).

43. J. H. Nobbs and I. M. Ward, Polymer Photophysics (D. Phillips, ed.), Chapman and Hall, London, p. 159 (1985).

44. J. P. Jarry and L. Monnerie, J. Polym. Sci., Polym. Phys. Ed., 18: 1879 (1980).

45. G. A. George, Developments in Polymer Degradation, Vol. 3, (N. Grassie, ed.), Applied Science, London, p. 173 (1981).

46. V. A. Belyakov and R. F. Vassil'ev, Photochem. Photobiol., 11: 179 (1970).

47. R. E. Kellogg, J. Am. Chem. Soc., 91: 5433 (1969).

48. B. A. Lloyd, K. L. DeVries, and M. L. Williams, J. Polym. Sci. A-2, 10: 1415 (1972).

49. D. J. Carlsson, K. H. Chan, J. Durmis, and D. M. Wiles, J. Polym. Sci., Polym. Chem. Ed., 20: 575 (1982).

50. R. J. Fleming and J. Hagekyriakou, Radiat. Protect. Dosim., 8: 99 (1984).

51. R. J. Fleming, C. A. Legge, J. H. Ranicar, and S. Boronkay, J. Phys. E., 4: 286 (1971).

52. I. Boustead, J. Polym. Sci. A-2, 8: 143 (1970).

53. J. Pospisil, I. Chudacek, and E. Donth, Polym. Bull., 12: 149 (1984).

54. W. G. Herkstroeter, J. Poly,. Sci., Polym. Schem. Ed., 22: 2395 (1984).

55. L. Bokobza, E. Pajot-Augy, L. Monnerie, A. Castellan, and H. Bouas-Laurent, Polym. Photochem., 5: 191 (1984).

56. E. Pajot-Augy, L. Bokobza, L. Monnerie, A. Castellan, and H. Bouas-Laurent, Macromolecules, 17: 1490 (1984).

57. F. W. Wang, R. E. Lowry, and B. M. Fanconi, Polymer, 27: 1529 (1986).

58. R. O. Loutfy, Pure Appl. Chem., 58: 1238 (1986).

59. R. L. Levy, Polym. Mater. Sci. Eng., 50: 124 (1984).

60. S. F. Scarlata and J. A. Ors, Polym. Commun., 27: 41 (1986).

61. K. Horie, K. Morishita, and I. Mita, Macromolecules, 17: 1746 (1984).

62. K. R. Smit, R. Sakurovs, and K. P. Ghiggino, Eur. Polym. J., 19: 49 (1983).

63. A. C. Somersall, E. Dan, and J. E. Guillet, Macromolecules, 7: 233 (1974).

2
Polymer Analysis by Luminescence Spectroscopy.

NORMAN S. ALLEN   Manchester Polytechnic, Manchester, England

ERYL D. OWEN   University College, Cardiff, South Wales

I.   INTRODUCTION

Over the last 25 years luminescence (fluorescence and phosphorescence) spectroscopy has proved to be extremely useful as an analytical tool in polymer science.   For example, the technique has provided valuable information on the mechanisms of polymer oxidation/degradation and stabilization [1—4] and this area is discussed in some depth in Chapters 4 and 3.   The technique has also been found to be a useful analytical tool in polymer science with regard to information on polymer properties such as their macro- and supermolecular structure as well as the determination of molecular weight, viscosity, and permeability [1—16].   The first two have attracted significant interest in recent years since the technique can provide access to information not readily available by other methods. The latter, however, have not been put into use practically because other, more reliable methods of measurement are available.   Another, more recent application has been in the identification of polymer systems [1—4,15] and certain additives in polymers [1—4,17,18]. There is often a frequent call for the rapid identification of polymer systems, including additives, particularly by scientists and technologists interested in rival products of other companies and as a routine relative semiquantitative method of polymer quality control. In this respect the technique could be very useful. *Graphs, spectra.*

THE INSTITUTE OF METALS

In this chapter a number of representative samples of the more practicable uses of luminescence analysis are highlighted for commercial synthetic polymers.

II. MEASUREMENT TECHNIQUES FOR POLYMERS

The techniques of fluorescence and phosphorescence may be readily adapted to suit most commercial forms of polymer material for analysis. However, as will be shown later, the form and origin of the polymer material can influence the properties of the emission spectra, and, as for other analytical methods, this information should be obtained using your own equipment for standard polymer samples. Even luminescence spectrometers differ in wavelength response and unless corrected the results will vary. Because of their poor stability many commercial polymers have to be studied in the solid state. Commercially, this may involve handling powders, films, sheet, granules, and fibers.

Polymer samples in film form (or sheet) can be easily studied in a fluorescence cell (cuvette) by placing them at an angle of 45° to the beam of excitation light. The film is positioned such that the excitation light is directed away from the emission monochromator in order to minimize scattering. A modification of the fluorescence technique specifically for studying polymer films has been developed [19]. A modified experimental setup is shown in Figure 1 and can be applied to most modern fluorimeters, provided you do not mind drilling a hole in the lid of the sample compartment. It has many advantages, such as good sample intensity reproduction, since the

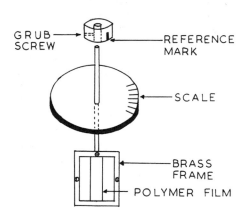

FIGURE 1 Experimental arrangement for measuring the fluorescence of polymer films. (Adapted from Ref. 19.)

angle of the polymer film can be kept constant. Using the metal (brass) frame (sandwich type) that I have drawn, the film can also be prevented from buckling.

Other polymer samples in the form of powders, granules, and fibers present problems but can be examined using the phosphorescence attachment. In older types of fluorimeters the rotating can has to be removed or lowered in order to observe fluorescence emission as well.

For phosphorescence measurements the Dewar arrangement, shown in Figure 2, is used as the cell. In this case the rotating can is put into position in order to eliminate any fluorescence emission and the sample has to be cooled with liquid nitrogen. Film or sheet samples cut into strip can be easily inserted into the silica tube or cell compartment of the Dewar. The Dewar is then turned until maximum emission is observed on the display or recorder. In modern instruments the rotating can is replaced by electronic gates.

In up-to-date instruments Dewars are no longer used. They are replaced by cooling blocks part of which is cut away to expose the silica sample tube (5 mm 0.10) to the excitation light. The powder, granules, or fibers can be easily inserted in the silica tube. Large polymer chip will have to be ground up and here a small coffee grinder and liquid nitrogen comes in very handy. For the older instruments the Dewar itself can be simply filled with chip material.

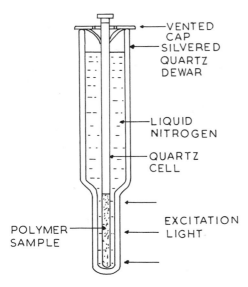

FIGURE 2 Experimental arrangement for the detection of phosphorescence from polymer samples.

III. ORIGIN OF POLYMER LUMINESCENCE

Somersall and Guillet have polymers classified into two main types
[9].   The first, termed type A, emit light through isolated impurity
chromophores situated as in-chain, side chain, or end chain groups.
The second, termed type B, emit light through chromophores present
in the repeat unit (or units) that form the backbone structure of
the polymer.   Typical type A and B polymers may be represented
by the general Structures I to VI.

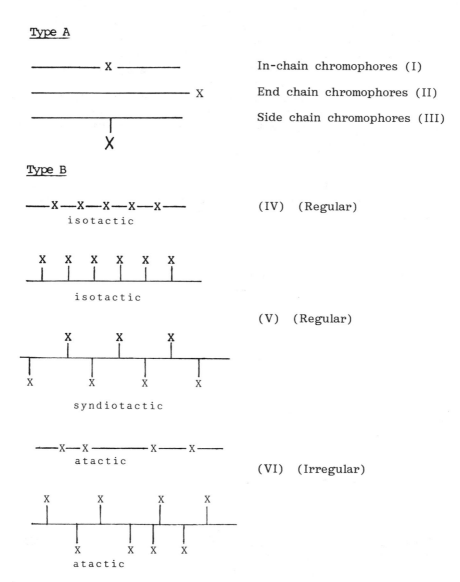

Type A

In-chain chromophores (I)

End chain chromophores (II)

Side chain chromophores (III)

Type B

(IV) (Regular)

isotactic

isotactic

(V) (Regular)

syndiotactic

atactic

(VI) (Irregular)

atactic

Commercial type A polymers include the polyolefins, synthetic rubbers, aliphatic polyamides, polyurethanes, polyvinyl halides, aliphatic polyesters, polystyrenes, polyacrylics, and polyacetals. An early study of the luminescence properties of type A polymers indicated the fluorescence and phosphorescence emissions may have a common origin irrespective of the chemical structure of the polymer [20]. At that particular time luminescence spectrometers were lacking in many respects, particularly in terms of correction and resolution, and consequently many of the emission spectra tended to be rather broad and structureless. However, despite this problem recent work with improved instrumentation does tend to support this conclusion for a number of polymers. Typical examples of luminescent chromophores present in type A polymers are carbonyl and α, β-unsaturated carbonyl groups. These have been found to be responsible for the fluorescence and phosphorescence emissions from polyolefins [21–24], aliphatic polyamides [25–29], synthetic rubbers [30,31], polyacetals and polyvinylchloride [34–36]. The α, β-unsaturated carbonyls, if situated in the main chain of the polymer, could have the following general structures:

$$\sim CH=CH-\underset{\underset{O}{\|}}{C}-R \qquad (VII)$$

$$\sim CH=CH-\underset{\underset{O}{\|}}{C}-CH=CH-R \qquad (VIII)$$

$$\sim CH=CH-CH=CH-\underset{\underset{O}{\|}}{C}-R \qquad (IX)$$

where R = end of chain or a continuous chain. The luminescence properties of these types of structures are very sensitive to both environmental and structural effects [29]. Simple aliphatic molecules of this type do not fluoresce or phosphoresce due to facile deactivation within both the singlet and triplet manifolds attributed to free roation about the double bonds. Thus, only cyclic α, β-unsaturated carbonyls have ever been reported to luminesce [29]. In a polymer network this cyclization would be effective through intramolecular chain coiling.

However, the exact nature of the luminescence from these types of polymers has been the subject of some controversy. In the case of the polyolefins, such as low-density polyethylene and polypropylene, early studies indicated that the presence of polynuclear aromatic hydrocarbons such as naphthalene were responsible for the fluorescence emissions [37–40]. This conclusion was based on the observation that some of the fluorescence is extractable from the

polymer by n-hexane and that it subsequently regenerates in air
over a given period of time. This was recently supported in similar
studies, although the authors did not commit themselves to an exact
identification of the chromophores [41,42]. Other workers, however,
have shown that the fluorescence and phosphorescence emissions
from polyolefins are associated with some oxidative process and are
due to the presence of α, β-unsaturated carbonyls of the enone
(VII) (en-al) and dienone (or-al) types (VIII/IX). Using corrected
excitation and emission spectra the fluorescence emission from all
four polyolefins, namely, polypropylene, low/high-density polyethylenes, and poly(4-methylpent-1-ene), has been found not to match
that of naphthalene under the same conditions [24]. An example of
this analysis is shown in Figure 3 where it is evident from the
excitation spectra that naphthalene does not resemble the n-hexane
extracted species from polypropylene [24]. Furthermore, the results
in Figure 4 demonstrate that the fluorescence regeneration process,
after treatment of polypropylene films with n-hexane, is an oxidative
one since the regeneration is clearly much faster in a pure oxygen

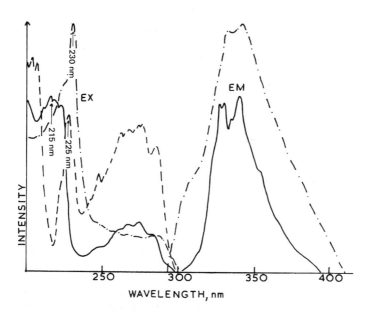

FIGURE 3 Fluorescence excitation and emission spectra of (———)
naphthalene in n-hexane (5 x 10⁻⁶ M) and (—·—·—) polypropylene
n-hexane extracts (1 g powder/40 cm³ n-hexane). Sensitivity x4.)
(- - -) Naphthalene excitation spectrum at 10⁻⁵ M in n-hexane. (Re-
proCuced from Ref. 24 with permission from Elsevier Applied Science
Publishers, London.)

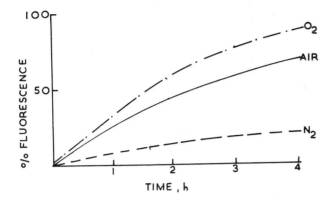

FIGURE 4 Rate of regeneration of fluorescence emission from poly-propylene film in (——) air, (- - - -) N_2, and (-·-·-) O_2 atmospheres. (Ex, 280 nm; Em, 340 nm). (Reproduced from Ref. 24 with permission from Elsevier Applied Science Publishers, London.)

atmosphere than in air. Other workers [43] have studied the luminescence from polyethylene in electric fields and found that the latter enhances the emission intensity. Doping the polymer with polynuclear aromatics did not result in any change thus implying that the luminescent species in polyolefins are structural rather than just free chemical entities.

Aliphatic polyamides such as nylon 6,6 are even more complex. This polymer, for example, contains at least two major types of fluorescent species as shown by the spectra in Figure 5 [29]. The fluorescence emission at 326 nm is similar to that of the extractable species in polyolefins but there is another chromophoric which emits at 390 and 420 mm and has been associated with α-ketoimides of the Structure X,

$$\sim CH_2-\underset{\underset{O}{\|}}{C}-\underset{\underset{O}{\|}}{C}-\underset{\underset{H}{|}}{N}-\underset{\underset{O}{\|}}{C}-CH_2 \sim$$

(X)

introduced into the polymer during manufacture and thermal process-ing [44].

The phosphorescence emission from nylon 6,6 is even more com-plex and is evidently due to the presence of a variety of types of cyclic saturated and α, β-unsaturated carbonyl compounds. This is demonstrated by the results in Figure 6 [29]. In fact, a recent detailed GC/MS and luminescence study [29] showed that all the

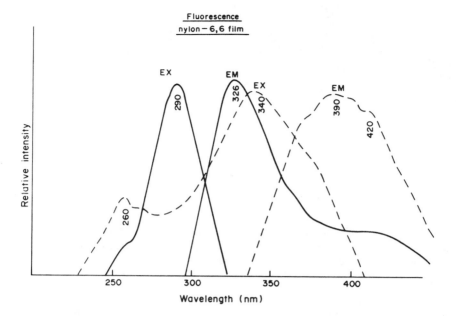

FIGURE 5   Fluorescence excitation (Ex) and emission (Em) spectra
of nylon-6,6 film (100 μm thick).   (Reproduced from Ref. 29 with
permission from Pergamon Press, Oxford.)

fluorescent and phosphorescent species in nylon 6,6 originate from
the aldol condensation reaction of cyclopentanone.   The latter is a
thermal degradation product of the polymer and was easily identifiable
in 2-propanol extracts of the polymer.   A full list of the products
which were identified are shown in Table 1.   Products (II) and
(VIII) in the table are believed to be associated with the fluorescence
emission at 326 nm from the film (Figure 5).   Unsaturated carbonyl
groups have also been found to be produced as a consequence of
thermal oxidation in polybutadiene [45] and polyvinylchloride (PVC)
[35], and spectral examples are shown in Figures 7 and 8, respectively.   In the latter case some early workers attributed the fluorescence to the presence of polyconjugation [34] while more recent
studies suggest that conjugation with carbonylic groups is essential
for emission to occur [35,36].

Other commercially important members of the type A classification
are the polyurethanes.   In one study on an MDI-based polyurethane
(diphenylurethane-4,4'-diisocyanate) of the general structure XI,
the phosphorescence emission was associated with the presence of
a benzophenone-type chromophore XII (Figure 9) [46].

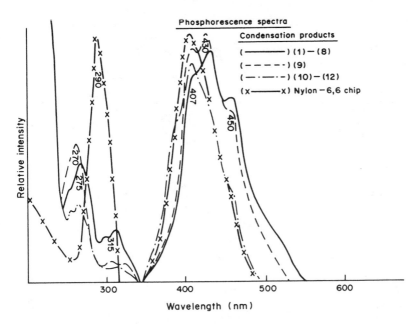

FIGURE 6 Phosphorescence excitation and emission spectra (77K) of aldol condensation products of cyclopentanone, TLC fractions (——) (1)–(8), (----) (9), and (–·–) (10)–(12) in 2-propanol and of nylon 6,6 chip (–x–). (Reproduced from Ref. 29 with permission from Pergamon Press, Oxford.)

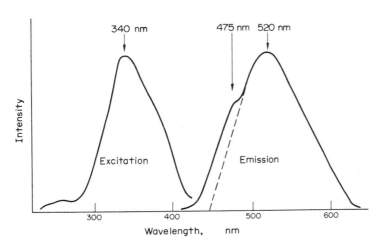

FIGURE 7 Phosphorescence excitation and emission spectra of thermally oxidized PBD. (Reproduced from Ref. 45 with permission from Pergamon Press, Oxford.)

TABLE 1 TLC-Isolated Products from the Aldol Condensation of
Cyclopentanone Analyzed by GC/MS

Source: Reproduced from Ref. 29 with permission from Pergamon
Press, Oxford.

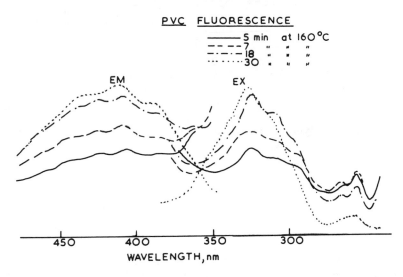

FIGURE 8 Fluorescence excitation (Ex) and emission (Em) spectra of PVC film after compression molding in air. (Reproduced from Ref. 35 with permission from Elsevier Applied Science Publishers, London.)

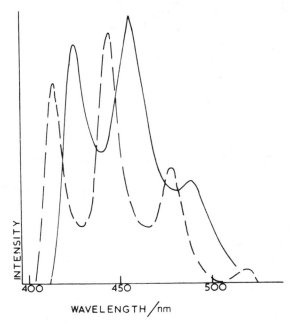

FIGURE 9 Phosphorescence emission spectra of an MDI-based poly-urethane film (——) and benzophenone (---) in 2-propanol (77K). Excitation = 325 nm. (Reproduced from Ref. 46 with permission from John Wiley and Sons, New York.)

$$\left(R\text{O}\cdot\text{OC HN}-\!\!\!\left\langle\!\!\!\bigcirc\!\!\!\right\rangle\!\!\!-\text{CH}_2\!\!\!\left\langle\!\!\!\bigcirc\!\!\!\right\rangle\!\!\!-\text{NHCO}\cdot\text{O} \right)_n$$

(XI)

where R = aliphatic polyester or glycol residue.

$$\sim\!\text{HN}-\!\!\!\left\langle\!\!\!\bigcirc\!\!\!\right\rangle\!\!\!-\underset{\underset{\text{O}}{\|}}{\text{C}}-\!\!\!\left\langle\!\!\!\bigcirc\!\!\!\right\rangle\!\!\!-\text{NH}\!\sim$$

(XII)

Other workers, however, believe that the emission in Figure 9 is due to a triplet excimer but as yet confirmation of this has not appeared [47].

In the use of natural polymers, wool has received much attention [48–51]. The fluorescence has been assigned to N-fromylkynurenine units formed from the oxidation of tryptophenyl residues while the phosphorescence emission is associated with the latter [50,51]. Figure 10 shows that the phosphorescence emission from wool matches that of tryptophan doped in polyvinyl alcohol [51].

Commercial type B polymers include the aromatic polyesters, polyethersulfones, aromatic polyamides, some polysulfides, polycarbonates, poly(phenylene oxides), polyimides, and many aromatic resins (e.g., phenolic resin). Apart from the normal fluorescence and phosphorescence emissions from unit structures of these polymers, many of them also exhibit other interesting photophysical phenomena. Owing to the high degree of conjugation in many of these polymers, various types of interactions and energy transfer processes are possible and these can give valuable information on chain conformations. Basically, there are two possible types of energy transfer, namely, intermolecular and intramolecular [6,9,16[.

Intermolecular energy transfer may involve the transfer of energy from small-molecule chromophores to those on a polymer chain or, alternatively, from the chromophores on a polymer chain to a small-molecule chromophore. Intermolecular energy transfer may also occur between two polymer molecules.

Intramolecular energy transfer, on the other hand, occurs within the same polymer molecule and in principle can be considered as energy transfer between different parts or segments of the same molecule. There are basically two types of intramolecular energy transfer in polymers. These are (1) between near-adjacent conjugated chromophores (XIII) and between nonadjacent or conjugated chromophores (XIV), e.g.

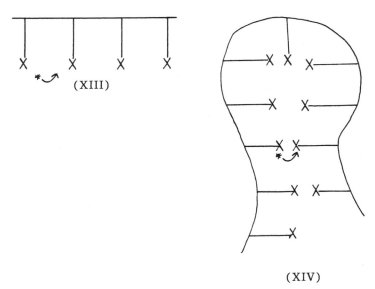

(XIII)

(XIV)

In the former process (XIII) energy migration (or energy hopping)
along the polymer chain can often occur over very long distances.
The migrating "exciton" will move along the polymer chain until it
is captured by an impurity or trap (e.g., physical defect in the
polymer). When two migrating triplet excitons are in the same
polymer chain, mutual annihilation may occur and delayed fluorescence
is observed. This omission corresponds with the normal fluorescence
spectrum of the polymer but is long-lived since it originates from a
triplet state. This process occurs in the more esoteric polymer
materials like the poly(vinylnaphthalenes) and poly(vinylcarbazoles)
[11,16].

In the second process (XIV) excimer formation is predominant
and has been observed in a number of polymers, e.g., polystyrenes,
poly(vinylcarbazoles), poly(vinylnaphthalenes), and poly(vinyltoluenes)
[11]. An excimer is an excited associated complex formed between
a species in the excited singlet state and a similar ground state
species, thus [11,16]:

$$A* + A \longrightarrow (AA*) \longrightarrow A + A + h\nu$$

Excimer fluorescence occurs at longer wavelength than that of normal
fluorescence and is very dependent on the concentration of the
species involved. Time-resolved fluorescence spectroscopy has been
widely used for the study of the time evolution of excited species in
polymer systems [52–54]. In this method by the use of a time-gating
system the normal and excimer fluorescence of polymers may be
observed separately after excitation.

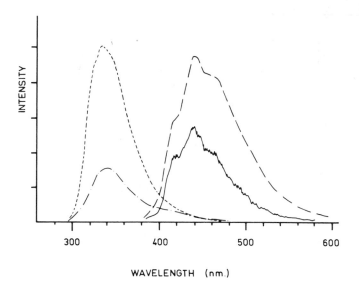

FIGURE 10   Luminescence emissions at room temperature: (−·−·)
total emission from wool keratin; (——) phosphorescence emission from
keratin; (····) total emission from tryptophan-PVA film; (---) phosphorescence emission from tryptophan-PVA film.   Spectra recorded
at different gains.   (Reproduced from Ref. 51 with permission from
Elsevier Sequoia, Switzerland.)

The fluorescence of one type B polymer, namely, poly(ethylene-
2,6-naphthalate), is associated with the monomer unit but is slightly
red-shifted due to excimer formation [55,56] (Figure 11).   The
fluorescence emission from poly(ethyleneterephthalate) has been
attributed to an associated ground state complex formed between the
terephthalate units, whereas the phosphorescence emission originates
from the monomeric unit structure of the polymer [57].   Poly(p-
xylylene) is an interesting case and one which exemplifies the complex nature of the fluorescence emission from highly conjugated
aromatic polymers [58].   This particular polymer is made from the
dimer 2,2-paracyclophane of the Structure XV to give a polymeric
Structure XVI.

(XV)   (XVI)

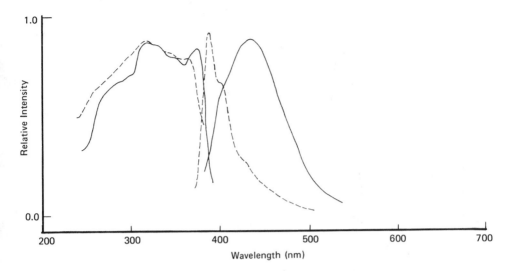

FIGURE 11 Comparison of the corrected fluorescence excitation and emission spectra of polyethylene-2,6-naphthalate film (——) with that of 2,6-dimethylnaphthalene crystals (---) at 298K. Excitation scan (——), Em 430 nm; emission scan (——), Ex 375 nm; excitation scan (---), Em 387 nm; emission scan (---), Ex 365 nm. (Reproduced from Ref. 56 with permission from John Wiley and Sons, New York.)

The fluorescence emission of poly(p-xylylene) shown in Figure 12 is composed of three components: monomer emission, excimer emission, and emission from an associated ground state complex. For bisphenol A-based epoxy resins the phosphorescence has been associated with monomer, associated ground state dimer, and a triplet excimer [59], while in the case of poly(oxy(2,6-dimethyl)-1,4-phenylene) the phosphorescence is due solely to the monomer unit whereas the fluorescence is due to monomer unit as well as xanthone and quininoid oxidation products [60]. The fluorescence of polyaramids has also been studied and is assumed to be due to monomer emission, although more in-depth studies in future may indicate other interesting features [61,62].

Emission from an associated ground state dimer occurs in the following simple reaction:

$$A \ldots A \xrightarrow{h\nu} (A \ldots A^*) \longrightarrow A \ldots A + h\nu'$$

Here the complex (A...A) is stable in the ground and excited states, and consequently both the excitation and emission wavelengths occur at longer wavelengths than those of the macromeric unit of the polymer.

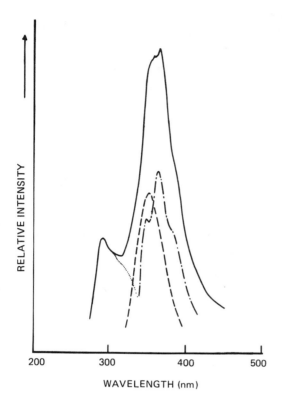

FIGURE 12 Fluorescence emission spectrum of poly($\underline{p}$-xylylene) (PPX) film, composed of three distinct emissions due to monomer emission (···), excimer emission (---), and emission from a ground state-associated complex (−·−·−). (Reproduced from Ref. 58.)

Another commercially important polymer that has been extensively studied is polystyrene. Evidence from the literature indicates the fluorescence and phosphorescence spectra of commercial polystyrene are very complex and have in fact attracted some in controversy in recent years [37,38,63−68]. Discussion on the latter point would be beyond the aims of this chapter but the main features of the fluorescence classify this polymer as one of type B, whereas the presence of impurities classify the polymer as type A. The fluorescence emission spectrum of polystyrene is often very dependent on the commercial source and purity of the polymer [66−69]. All commercial samples of this polymer exhibit fluorescence emission at about 310 nm due to residual styrene monomer [64,68]. However, thin films of "pure" polymer show only excimer emission on excitation at the absorption maximum of the polymer (λ_{max} = 250 nm) [63,68].

In solution both monomer and excimer emission is observed [63,65]. The intensity of the excimer emission increases with an increase in the concentration of the polymer and a red shift in the wavelength maximum also occurs [65]. This effect has been attributed to intramolecular quenching of adjacent phenyl chromophores [65].

Commercial polystyrene also exhibits an "anomalous" fluorescence with emission maxima at 338, 354, and 372 nm on an excitation with light of wavelengths greater than 300 nm [68]. This emission has been attributed to the presence of trans-stilbene linkages (XVII).

$$\sim CH_2 \; \underset{\displaystyle \bigcirc}{C} = \underset{\displaystyle \bigcirc}{C} - CH_2 \sim \qquad\qquad \sim CH_2 - \underset{\displaystyle \bigcirc}{C} {=} O$$

(XVII)   (XVIII)

The phosphorescence emission, on the other hand, has been assigned to the presence of acetophenone-type terminal groups of the Structure XVIII. The luminescence properties of several types of both A and B polymers are collated in Table 2, which has been updated by the authors from the previously published version [69].

IV. IDENTIFICATION OF COMMERCIAL POLYMERS

The luminescence properties of polymers can be useful for their identification, particularly if they are used in conjunction with other techniques such as infrared spectroscopy [69]. The use of luminescence spectroscopy as an analytical technique for polymer identification involves the measurement of the following properties:

1. The fluorescence emission spectrum, which is obtained by exciting the polymer with light of any wavelength that is capable of producing the emission spectrum. The most satisfactory spectrum for definitive purposes, however, is normally recorded at the wavelength maximum (λ_{max}) of the excitation spectrum.
2. The fluorescence excitation spectrum, which is obtained by recording the variation in intensity of the emission spectrum (λ_{max}) as a function of the excitation wavelength. The corrected excitation spectrum should closely match the absorption spectrum of the chromophore provided it is corrected.
3. The phosphorescence emission spectrum, which is obtained by using the same method as that used for the fluorescence emission spectrum.

TABLE 2 Fluorescence and Phosphorescence Properties of Polymers

Polymer	Form	Excitation (nm)	Emission (nm)	Mean lifetime (sec)	Chromophore
			Fluorescence		
Poly(ethylene-terephthalate)	Chip	320,344,357(s)	370,389,405	—	Polymer (dimer)
	Film	344,357	370,389,405	—	Polymer (dimer)
	Fiber	344,357	370,389,405	—	Polymer (dimer)
Poly(ethylene-2,6-naphthalate)	Film	375	435	—	Polymer (excimer)
Polyurethane, MDI-based	Film	372	420	—	Unknown
Nylon 6,6	Chip	357	417	4ns	α-ketoimides
	Fiber	357	417	4ns	α-ketoimides
Nylon 6	Chip	335	390	—	α-ketoimides
Nylon 6,10	Chip	345,355	395,410	—	α-ketoimides
Nylon 11	Chip	327,340	375(s),395(s),385(s)	—	α-ketoimides
Nylon 12	Chip	410	450	—	α-ketoimides
Polyvinylfluoride	Film	290	350	—	α,β-Unsaturated carbonyl (enone or -al)
Polyvinylfluoride (heated)	Film	325	410	—	α,β-Unsaturated carbonyl (dienone- or -al)
Poly(oxy(2,6-dimethyl)-1,4-phenylene)	Soln	290	310	—	Polymer
	Film	495,515	550,565	—	xanthanoid (oxidation)

Sample	Form				Assignment
Wool	Fiber	290	345	—	Aromatic tryptophyl residues (N-formyl-kynurenine)
Poly(m̲-benzamide)	Film	405	500	—	Polymer
Poly(m̲-phenylene isophthalamide)	Film	350,410(s)	465	—	Polymer
Poly(m̲-phenylene terephthalamide)	Film	375	470	—	Polymer
Bisphenol A-based epoxy resins	Soln	350	424	—	Impurity in amine
Polytetrafluoroethylene	Film	328	350	—	Unknown
Polyvinyl alcohol	Film	258(s),295,330	360,370(s)	—	Unknown
Poly(butylene terephthalate)	Soln	255,290	324	—	Polymer
	Chip	332	400,420,450	—	2-hydroxyterephthalic acid units
Polyethylene (low-density)	Powder	230,265(s),300	335(s),350	—	Cyclic
	Film	230,273	295(s),310,329(s),354(s),370(s)	—	
Polyethylene (high-density)	Film	230,265(s),290	295(s),312(s),330(s),344(s),358	—	α,β-unsaturated carbonyl groups of the enone (or -al) type
Poly(ethylenevinyl-acetate)	Film	230,265(s),290	312(s),330,344(s),358(s)	—	
Polypropylene	Film	230,285	309(s),320	—	
Poly(4-methyl-pent-1-ene)	Film	230,285	310,330	—	

TABLE 2 Continued

Polymer	Form	Excitation (nm)	Emission (nm)	Mean lifetime (sec)	Chromophore
Polystyrene	Chip	318,330	336,354,368(s)		trans-stilbene
Polyethersulfone	Film	320	360		Polymer (excimer)
Poly(p-xylylene)	Film	265	296	—	Monomer
		265	365	—	Excimer
		307,327,339	351(s),370,390(s)	—	Polymer (dimer)
Polyoxymethylene		210,285	320	—	α,β-unsaturated carbonyl groups
			Phosphorescence		
Polyethylene-terephthalate	Chip	280,318,351	425(s),460	0.5	Polymer
	Fiber	284,310	425(s),477	0.7	Polymer
Poly(ethylene-2, 6-naphthalate)	Film	375	580	—	Polymer
Polyurethane MD1-based	Film	320	423,455,489	0.02	Benzophenone type
Nylon 6,6	Chip	296	400	2.1	
	Fiber	296	430	1.3	
Nylon 6	Chip	282	390(s),420,455(s)	1.7,1.6,1.1	Cyclic
Nylon 6,10	Chip	300	430	0.70	α,β-unsaturated carbonyl groups of dienone (or -al) type

Material	Form				
Nylon 11	Chip	296(s),273(s)	423,450(s)	1.0,0.88	
Nylon 12	Chip	268,286(s)	363(s),410	1.0	
Polyvinylchloride	Film	284	440	0.30	Carbonyl
Polytetrafluoro-ethylene	Film	260–280+	450	0.40	Carbonyl
Polyvinyl alcohol	Film	260–280+	436	0.40	Unknown
Polyethylene* (low-density)	Powder	273,280	367,381,391,405,416	2.30	Benzoic acid
Polyethylene (low-density)	Film	278,280	420	0.60	Benzoic acid
Polyethylene# (low-density)	Chip	283,331	370(s),435,455	2.15	
Polyethylene-vinyl acetate	Film	280,327	455	0.35	Cyclic α,β-unsaturated carbonyl groups of dienone (or -al) type
Polypropylene	Film	270,290,330	420,445,480,510(s)	0.5 −1.2	
Poly(4-methylpent-1-ene)	Film	273,330	430	0.86	
Polyethylene (high-density)	Film	275	450	0.35	
Polystyrene	Film	290(s),300,336(s)	398,425,456,492	0.008	Acetophenone end groups
Polyethersulfone	Film	320	450	0.05	Polymer
Polyoxymethylene	Film	290,320(s)	415	1.0	α,β-Unsaturated carbonyl species
Wool		320	405,425,450	4.76	Tryptophone residues

TABLE 2 Continued

Polymer	Form	Excitation (nm)	Emission (nm)	Mean lifetime (sec)	Chromophore
Bisphenol A epoxy resin	Film	275,350	460	—	Monomer/dimer and triplet excimer
Poly(oxy(2,6-dimethyl)-1,4-phenylene)		325	460	0.04	Polymer
Poly(butylene terephthalate)	Fiber	305	450	1.2	Polymer
Poly(p-xylylene)		280	402,430,447,461,478,493	—	Dimer

(s) = Shoulder.

MO1 = Diphenylmethane-4,4-diisocyanate.

+ = Broad and structureless spectrum.

* = Prepared using a benzoyl-based catalyst.

= Prepared using oxygen or aliphatic peroxide catalyst.

Source: Adapted from Ref. 69 with permission from the Royal Society of Chemistry, London.

4. The phosphorescence excitation spectrum, which is obtained by using the same method as that used for the fluorescence excitation spectrum.
5. The phosphorescence lifetime, which is determined by coupling the emission signal to either a chart recorder or, better still, an oscilloscope and measuring the decay of the emission signal with time while the excitation light is cut off. Emission lifetimes are taken as the time for the emission to decay to 1/e of its original intensity. In more modern instruments the emission signal intensity can be altered by changing the gating time and calculating the lifetime on a data station using an appropriate obey file from floppy disk.

If it is found that the wavelength maximum (λ_{max}) of either the fluorescence or phosphorescence spectrum varies with the excitation wavelength, the presence of more than one type of chromophore is indicated and is necessary to examine the polymer over a wide range of excitation wavelengths in order to achieve its complete identification. The above luminescence characteristics of a number of commercial polymers are shown in Table 2 together with the identities of the chromophores that are believed to be responsible for the various emissions.

All the polymers shown in the table tend to fluoresce in the wavelength range 300 to 450 nm and phosphoresce in the region 400 to 600 nm. In certain cases the spectra are highly structured and this can assist further in the identification of the polymer. The highly structured phosphorescence excitation and emission spectra of poly(ethyleneterephthalate), wool, polystyrene, and polyethylene shown in Figure 13 are good examples to illustrate this feature. The manufacturing history of a polymer can in some cases influence the nature of the luminescence. This effect is demonstrated in Table 2 for polyethylene, where the phosphorescence emission of polymer prepared using a benzoyl peroxide-type catalyst differs from that of polymer prepared using oxygen as a catalyst [22]. The form of this polymer is also an important factor to be considered as shown by the phosphorescence emission data in Table 2 for film and powder materials.

The advantages and disadvantages of luminescence spectroscopy for identifying polymers are as follows:

Advantages:

1. The technique is rapid and nondestructive.
2. It requires no tedious sample preparation.
3. It can be used to analyze polymers in powder, chip, film, fiber, or sheet form. Normally, the amounts of polymer required are in the range 10 to 50 mg for powders and film and up to 100 mg for fibers and chip.

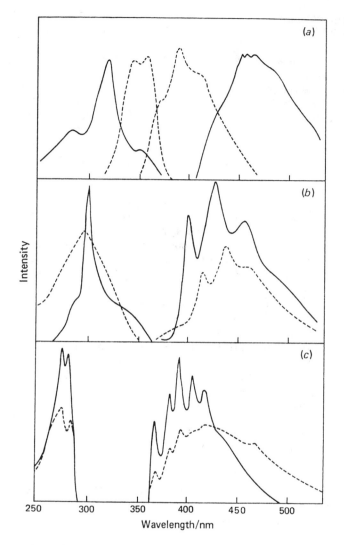

FIGURE 13   (a) Fluorescence (broken line) and phosphorescence
(full line) excitation and emission spectra of polyethyleneterephthalate
chip; (b) phosphorescence excitation and emission spectra of polystyrene chip (full line) and wool fiber (broken line); and (c) phosphorescence excitation and emission spectra of polyethtlene powder
(full line) and film (broken line).   (Reproduced from Ref. 59.)

4. It is a highly sensitive technique. Concentrations of 10^{-8} mol/ liter of aromatic hydrocarbons can be detected by fluorescence spectroscopy.
5. It can be used to characterize polymers within a particular class, e.g., the nylon polymers (see Table 2). This advantage would be useful in conjunction with other techniques such as infrared spectroscopy.

Disadvantages:

1. The technique gives no information on those polymers which are nonluminescent or only weakly so. However, this problem can sometimes be overcome by thermal oxidation of the material under controlled conditions. The nonvolatile oxidation products may be luminescent and often characteristic for that particular polymer [1,2].
2. Some light stabilizers reduce the intensity of the luminescence from polymers [70] and, if present, they must be removed by solvent extraction. Similarly, other additives such as antioxidants will almost certainly be present and in many cases will not interfere to any significant extent but may themselves be identified by luminescence analysis either after extraction or in situ.
3. The presence of pigments may provide difficulties.

Some pigments luminesce and can be analyzed accordingly.
It should be noted that since the number of polymers listed in Table 2 is by no means complete and because different instruments will give different wavelength values for the spectra, it is certainly recommended that the analyst constructs a table of data (or spectra) by using the available equipment. When compiling such a table of data, the form of the polymer should be noted as this will help eliminate any differences due to processing history and morphology.

V. ANALYSIS OF ADDITIVES IN POLYMERS

Many antioxidants and light stabilizers exhibit their own characteristic fluorescence and/or phosphorescence emissions and may therefore be analyzed after solvent extraction from the polymer. A number of the more recent hindered piperidine light stabilizers, however, do not luminesce and will not interfere with the emission from the polymer. The fluorescence and phosphorescence properties of a number of antioxidants and light stabilizers are given in Table 3 [18]. The detection limits of the additives vary quite markedly and this could be a problem, particularly for commercial systems containing mixtures of additives. In most commercial polymers an antioxidant and ultraviolet stabilizer mixture is used, and these may have to be separated using thin-layer chromatography (TLC) if their emission spectra

TABLE 3   Luminescence Characteristics of Polymer Additives

No.	Trade name	Chemical composition	Room temp. fluorescence ex λ_{max}	em λ_{max}
1.	Topanol A	2,4-Dimehtyl-6-t-butylphenol	—	—
2.	Topanol OC	2,6-Di-t-butyl-4-methylphenol	255	318[a]
3.	Tenox BHA	Mixture of 2- and 3-t-butyl-4-hydroxyanisole	312,255	380,335
4.	Binox M	Bis(3,5-di-t-butyl-4-hydroxyphenyl)methane	—	—
5.	Ionox 330	1,3,5-Trimethyl-2,4,6-tris-(3,5-di-t-butyl-4-hydroxybenzyl)benzene	295	335[a]
6.	Nonox WSP	Bis(2-hydroxy-3-α-methylcyclohexyl-5-methylphenyl)methane	372	440[a]
7.	Nonox WSL	2,4-Dimethyl-6-α-methylcyclohexylphenol	372	464[a]
8.	Nonox DCP	2,2-Bis(3-methyl-4-hydroxyphenyl)propane	310,370	390,460[a]
9.	Calco 2246	Bis(2-hydroxy-3-t-butyl-5-methylphenyl)methane	325	375[a]
10.	Topanol CA	1,1,3-Tris(2-methyl-4-hydroxy-5-t-butylpehnyl) methane	255,318	318,408[a]
11.	Santonox R	Bis(2-methyl-4-hydroxy-5-t-butylphenyl)sulfide	360	410[a]
12.	Topanol TP	Bis(2-hydroxy-3,5-di-t-butyl-6-methylphenyl)sulfide	—	—
13.	Suconox 18	N-Stearoyl-p-aminophenol	318,345	380[a]
14.	Naugawhite	Bis(2-hydroxy-3-nonyl-5-methylphenyl)methane	366,312	446,390[a]
15.	Agerite Superlite		345	405
16.	Voidox 100%	2,6-Di-t-butyl-4-methyl phenol+sorbitan/fatty acid compound	266	318[a]
17.	Irganox 1010	Pentaerythritol-tetra-β-(3,5-di-t-butyl-4-hydroxyphenyl)-propionate	300	350[a]

Low-temp. luminescence		Phosphorescence		Phosphorescence lifetime (sec)	Phosphorescence detection limit (ppm)
ex λ_{max}	em λ_{max}	ex λ_{max}	em λ_{max}		
289	425	285	420	0.5	1.0
—	—	—	—	—	—
299	420	295	420	1.80	0.06
—	—	—	—	—	—
—	—	—	—	—	—
290	415	290	412	1.53	0.1
296	430	282	415	0.5	1.0
290	410	285	410	1.9	0.05
282	410	285	410	1.56	0.2
280	405	285	405	0.70	0.07
300	430	305	430	0.035[e]	0.07
295	428	300	426	0.033[e]	0.1
300	415	300	415	1.0	0.12
285	408	285	410	1.60	0.08
290	410	290	420	1.85	0.06
295	420	292	435[b]	2.5	>10
328	380	—	—	—	—

TABLE 3   Continued

No.	Trade name	Chemical composition	Room temp. fluorescence ex λ_{max}	Room temp. fluorescence em λ_{max}
18.	Irganox 1076	n-Octadecyl-β-(3,5-di-t-butyl-4-hydroxyphenyl)-propionate	376	430
19.	Irganox 1093	Di-n-octadecyl-3,5-di-t-butyl-4-hydroxybenzylphosphonate	315	375[a]
20.	Polygard	Tris(nonylphenyl)phosphite	356	422[a]
21.	Nonox CI	N,N'-Di-β-naphthyl-p-phenylenediamine	392	490
22.	DLTDP	Dilaurylthiodipropionate	—	—
23.	Salol	Phenylsalicylate	340	464[a]
24.	Cyasorb UV9	2-Hydroxy-4-methoxy-benzophenone	—	—
25.	Cyasorb UV531	2-Hydroxy-4-n-octoxybenzo-phenone	—	—
26.	Uvinol 400	2,4-Dihydroxybenzophenone	—	—
27.	Cyasorb UV24	2,2'Dihydroxy-4-methoxybenzo-phenone	—	—
28.	Tinuvin P	2-(2'-Hydroxy-5'-methylphenyl)-benzotriazole	—	—
29.	Tinuvin 326	2-(2'-Hydroxy-5'-t-butylphenyl)-3-chlorobenzotriazole	—	—

[a]Denotes weak room temperature fluorescence.
[b]Denotes weak phosphorescence at 77°K, only approximate estimates of life-time made.
[e]Life-time measurement reproducible within ± 10%.
Source: Reproduced from Ref. 18 with permission from Elsevier Sequoia, Switzerland.

cannot be resolved. If available, one very useful analytical facility here would be a front face accessory for the fluorimeter for analyzing spots on TLC plates directly. The luminescence data quoted in Table 3 is out of date and again the analyst is recommended to set up a calibration chart. Some more recent quantitative luminescence

Low-temp. luminescence		Phosphorescence		Phosphorescence lifetime (sec)	Phosphorescence detection limit (ppm)
ex λ_{max}	em λ_{max}	ex λ_{max}	em λ_{max}		
—	—	—	—	—	—
—	—	—	—	—	—
280	400	282	400	1.75	0.1
386,434	430,455	382	516	0.90	0.02
—	—	—	—	—	—
320	452	302	415	0.45	0.1
300	450	300	450[b]	0.02	>10
300	450	300	450[b]	0.03	>10
—	—	—	—	—	—
305	430	308/360	430/455[b]	0.1	>10
315	396,480	315	480,515	0.85	0.03
—	—	—	—	—	—

data on four antioxidants is given in Table 4 [71] where it is seen that generally antioxidant fluorescence has a low quantum yield of 10^{-3} to 10^{-2}. In some cases the phosphorescence can be high. Often a polymer manufacturer producing large-scale amounts of stabilized material requires a frequent check on the polymer for the

additive level.   Direct studies on granules as they come off the
plant by luminescence analysis could be useful on a routine basis
provided the correct form of sampling and a reproducible sample setup is maintained.

VI.   ANALYSIS OF PIGMENTS IN POLYMERS

Certain types of pigments (like dyes) exhibit their own characteristic
luminescence spectra.   Of the many types of pigments available, white
pigments are the most widely used, particularly titanium and zinc
oxides.   The two crystallize modifications of titanium dioxide, anatase
and rutile, may be easily identified by their characteristic emissions
[72,75].   At low temperatures (77K) anatase exhibits a strong emission spectrum at 540 nm while rutile exhibits weak emission in the
infrared at 815 and 1015 nm ($\tau_{i/e} \cong 10^{-5}$ sec).   The excitation λ_{max}
of the anatase and rutile crystallize forms are 340 and 375 nm, respectively.   This difference in emission properties of the two forms
has also been associated with their marked difference in photoactivity
in a polymer [75].   The nature of the surface treatment, often
applied to titanium dioxide pigments for various commercial reasons,
also affects the intensity of the emissions from the pigments as well
as their photoactivity [73].   The manufacturing history of titanium
dioxide pigments can also be determined from their characteristic
emission spectra in the infrared.   For example, at low temperatures
(77K) rutile pigments manufactured by the "sulfate" process exhibit
emssion at 815 and 1015 nm, whereas those manufactured by the
"chloride" process exhibit emission at 1015 nm (Figure 14).   Anatase
pigments also give an emission at 1015 nm but its intensity is much
stronger than that from any of the rutile grades.   Zinc oxide has
also been reported to luminesce [76,77] and the emission intensity
is related to pigment conductivity and photoactivity.   The emission
properties of colored pigments has not been studied to date.

VII.   MOLECULAR STRUCTURE

Luminescence techniques have been widely used for the study of
molecular behavior in polymers [1—11,16,78].   For example, the
familiar glass transition temperature (T_g) is the point at which backbone segmental motion can occur.   Below T_g, polymer chain mobility
ceases to occur and other transitions are observed due to the rotation of groups either contained or substituted in the polymer backbone.   These molecular motions can affect the phosphorescence yield
due to an increase in the rate of nonradiative deactivation processes.
A typical plot of phosphorescence intensity against temperature for
various ethylene polymers is shown in Figure 15 [78].   Two main
transitions are observed for these polymers: one at 163K due to

TABLE 4 Fluorescence and Phosphorescence Emission λ_{max} and Quantum Yields (Φ_F, Φ_p) for Antioxidants in Solvents of Various Polarities

Antioxidants	n-Heptane				Ethyl acetate				n-Propane	
	λ_F	Φ_F	λ_p	Φ_p	λ_F	Φ_F	λ_p	Φ_p	λ_F	Φ_F
Cyasorb UV 2908	308	2×10^{-4}	425	0.32	315	2.2×10^{-4}	425	058	322	5.1×10^{-4}
Shell acid	312	3.6×10^{-4}	425	0.48	317	1.5×10^{-4}	425	0.61	317	5.1×10^{-4}
Topanol OC	304	3.9×10^{-3}	—	$<10^{-3}$	316	1.4×10^{-2}	—	$<10^{-3}$	316	1.7×10^{-2}
Irganox 1076	314	6.1×10^{-3}	—	$<10^{-3}$	316	1.0×10^{-2}	—	$<10^{-3}$	315	1.5×10^{-2}

Source: Reproduced from Ref. 71 with permission from Elsevier Applied Science Publishers Ltd., London.

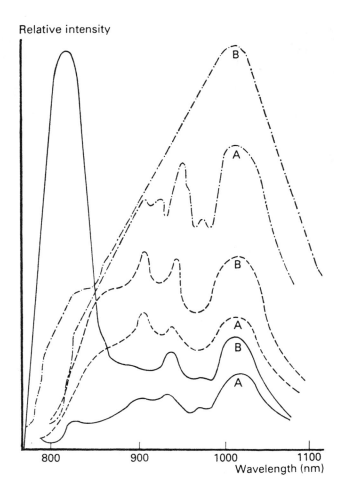

Relative intensity

Wavelength (nm)

FIGURE 14   Emission spectra of a "sulfate" processed rutile pigment
(——) (Sx100), a "chloride"-processed rutile pigment (---) (Sx100),
and a "sulfate"-processed anatase pigment (·—·-) (Sx100) (A),
(Sx30) (B), in polyethylene at room temperature (300K) and liquid
N_2 temperature (77K), respectively.   Excitation wavelength = 375 nm.
Sx = sample sensitivity.   Spectra were obtained using a Hitachi
Perkin-Elmer MPF-4 spectrofluorimeter.   (Reproduced from Ref. 73.)

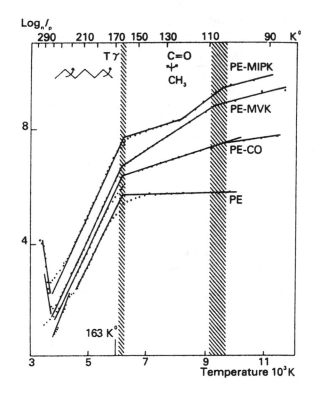

FIGURE 15 Arrhenius curves for polyethylene phosphorescence.
PE-MIPK, polyethylene co-methylisopropyl ketone; PE-MVK, poly-
ethylene-methylvinylketone; PE-CO, polyethylene-carbon monoxide.
(Reproduced from Ref. 78 with permission from the American Chemical
Society, Washington, D.C.)

crankshaft-type motion about the polymer chain; the other at 105K
due to rotation of the methyl group adjacent to the carbonyl.
Similar results have been obtained with other polyethers, particularly
naphthalene-labeled acrylics and labeled polystyrenes, and any form
of molecular motion tends to increase the probability of conversion
of electronic excitation energy to thermal energy, i.e., rotational
and vibrational.

The compatibility of polymer blends has attracted some interest
with the aid of luminescence analysis and utilizes measurements on
rates of intermolecular energy transfer and excimer formation on
chain conformation. Thus, the efficiency of both interchain singlet
energy transfer and the resulting disruption of excimer forming sites
depends on the degree of chain interpenetration and hence the mix-
ibility of the two polymers. In one particular study the compatibility

of poly(methyl methacrylate) with styrene-acrylonitrile copolymers
was determined by labeling the polymers with fluorescent anthracene
and carbazole groups, respectively, and then measuring the ratio of
the intensity of the emission from the carbazole in ralation to that
from the anthracene on excitation in the carbazole absorption band
(296 nm). The ratio of the emissions (Ic/Ia) are plotted against the
copolymer compositions in Figure 16 [79]. The compatibility range
is clearly indicated by the pronounced decrease in the Ic/Ia ratio
and this occurs between 30 to 40%, which is in agreement with the
results obtained by other methods.

Characterization of polymer blend compatibility has also been
carried out using excimer fluorescence. The effect of polymer
molecular weight on the kinetics of phase separation in polystyrene-
polyvinylether blends [80,81] has been determined by this method.
Since excimers require sandwich formation between the aromatic oxide
groups any chain interpenetration which occurs in a mixable blend
will reduce the available number of excimer-forming sites and hence
the excimer fluorescence intensity. Phase separation will therefore
give the greatest emission intensity. Numerous other studies on
this subject include blends of poly(2-vinylnaphthalene) with polystyrene and poly(methyl methacrylate) [82], anthracene-labeled

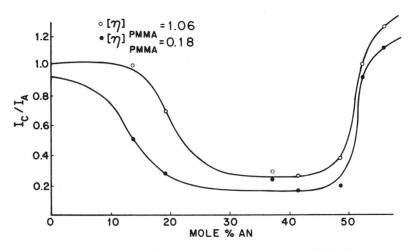

ENERGY TRANSFER IN PMMA/S-AN SYSTEM

FIGURE 16 Ratio of the emission intensity of donor and acceptor in
blends of chromophore labeled poly(methyl methacrylate) and styrene-
acrylonitrile copolymers as a function of the copolymer composition.
(Reproduced from Ref. 79 with permission from Pergamon Press,
Oxford.)

polystyrene and poly(α-methylstyrene) [83,84], and acrylonitrile-2-
($\underline{N}$-carbazoyl)ethyl methacrylate-styrene copolymer [85].   In all cases
the relationship between normal and excimer fluorescence emissions
are formed to provide more accurate information on blend imiscibility
than Differential Scanning Calorimetry (DSC).

     Fluorescence spectroscopy has also been useful for determining
the solubility parameters of soluble polymers at infinite dilution.
Polystyrene labeled with pyrene groups is one example [86].   Figure
17 shows that a plot of η_0 (Ie/Im) (excimer/monomer pyrene emission
intensities) versus the Hildebrand solubility parameter δ_H gives two
straight lines intersecting at $\delta_H = 9.1$.   This is the literature value
reported for polystyrene.   Molecular weight is also an important
factor here in enhancing excimer emission owing to the higher incidence of pyrene chromophores.   Critical concentrations of styrene
and methyl methacrylate copolymers in solution have also been
studied [87].

     Many studies of the behavior of luminescent additives dissolved
in solid polymers have yielded information about the polymer itself.
For example, the quenching of the phosphorescence of an aromatic

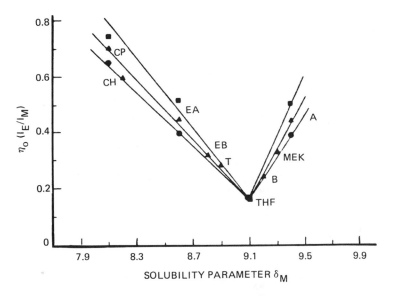

FIGURE 17   Plot of $\eta_0(I_E/I_M)$ vs δ_H for samples of polystyrene
labeled with pyrene groups: CP = cyclopentane, CH = cyclohexane,
EA = ethyl acetate, EB = ethylbenzene, T = toluene, A = acetone.
Molecular weight = 52, 200 ▲; 32, 900 ●; 13, 800 ■.   (Reproduced
from Ref. 86 with permission from the American Chemical Society,
Washington, D.C.)

additive by oxygen has formed the basis of a method for the measurement of the movement of the gas through the polymer [8]. The internal rheological behavior of certain polymers has been investigated through the polarization of the fluorescence of molecules added or bound to the polymer [6]. For example, a clear, slightly crosslinked rubber containing auramine, O, is fluorescent when stretched, but the intensity decreases on relaxation. The luminescence spectra of additives, such as anthracene, in polyethylene are highly sensitive to structural defects in the polymer and vary with the crystallinity of the region in which the emitting molecule is included [88]. While many of the latter methods applied to luminescence spectroscopy appear interesting, some have to date been adopted commercially.

VIII. MOLECULAR WEIGHT

The intensity of the fluorescence and phosphorescence emissions from various synthetic polymers has been related to polymer chain length with varying degrees of success [13,14]. For example, polystyrene fluorescence intensity decreases with an increase in its molecular weight (Figure 18). From the results shown in the figure the molec-

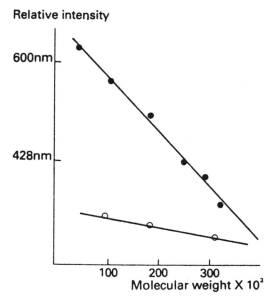

FIGURE 18 Fall in intensity of fluorescence of polystyrene at 428 and 600 nm with increase in molecular weight. (Reproduced from V. F. Gachkovskii, <u>Polym. Sci. USSR</u>, <u>2199</u> (1965), original source <u>Vysokomol. Soedin</u>, <u>A.7</u>:2009 (1965), with permission from Pergamon Press, Oxford.)

ular weight of an unknown sample of polymer may be determined by simply reducing its fluorescence intensity under certain conditions.

IX. CONCLUSIONS

Luminescence spectroscopy continues to show potential as a simple, versatile, and highly sensitive analytical method in a number of areas of polymer science and technology. The technique itself is constantly being improved, particularly on the measurement side, and while many of the above methods only appear to have been adopted academically, there is nevertheless a gradual growing awareness of its potential use in industrial laboratories where its advantages could have practical analytical applications.

REFERENCES

1. N. S. Allen and J. F. McKellar, Cham. Ind., (London), 907, (1978).

2. N. S. Allen, in UV Spectrometry Group Bulletin, 7: 51 (1979).

3. N. S. Allen, in Analysis of Polymer Systems (L. S. Bark and N. S. Allen, eds.), Elsevier, London, Chapt. 4, p. 79 (1982).

4. G. A. George, Pure Appl. Chem., 57: 945 (1985).

5. S. Czarnecki and M. Kryszewsiki, J. Polym. Sci., Polym. Chem. Ed., 1: 3067 (1963).

6. G. Oster and V. Nishijima, Fortschr. Hochpolym. Forsch., 3: 313 (1964).

7. E. I. Hannats and F. C. Underleitner, J. Phys. Chem., 69: 487 (1968).

8. G. Shaw, Trans. Farad. Soc., 63: 2181 (1967).

9. A. C. Somersall and J. E. Guillet, J. Macromol. Sci., Revs. Macromol. Chem., C13: 135 (1975).

10. A. M. North, Br. Polym. J., 7: 119 (1975).

11. J. E. Guillet, Polymer Photophysics and Photochemistry, Cambridge University Press, Cambridge, (1985).

12. J. F. McKellar and N. S. Allen, Photochemistry of Man-Made Polymers, Applied Science, London (1979).

13. V. F. Gachkovskii, Vysokomol Soedin, 17: 2007 (1975).

14. V. F. Gachkovskii, Adv. Mol. Relax. Proc., 5: 24 (1973).

15. S. W. Beavan, J. S. Hargeaves, and D. Phillips, Adv. Photochem., 11: 207 (1979).

16. D. A. Holden and J. E. Guillet, in Developments in Polymer Photochemistry, Vol. 1 (N. S. Allen, ed.), Applied Science, London, Chapt. 3 (1981).

17. H. V. Drushel and A. L. Sommers, Anal. Chem., 36: 836 (1964).

18. G. F. Kirkbright, R. Narayanqswarmy, and T. S. West, Anal. Chem. Acta, 52: 237 (1970).

19. J. F. McKellar and P. H. Turner, Fluorescence News, 7: 4 (1973).

20. V. F. Gachkovskii, Zhurnal Structumol. Khimii, 4: 424 (1963).

21. N. S. Allen, J. Hamer, and J. F. McKellar, J. Appl. Polym. Sci., 21: 2261 (1977).

22. N. S. Allen, J. Hamer, and J. F. McKellar, J. Appl. Polym. Sci., 21: 3147 (1977).

23. N. S. Allen and J. F. McKellar, J. Appl. Polym. Sci., 22: 625 (1978).

24. N. S. Allen, Polym. Deg. Stab., 6: 193 (1984).

25. N. S. Allen, J. F. McKellar, and D. Wilson, J. Photochem., 6: 337 (1976).

26. N. S. Allen, J. F. McKellar, and D. Wilson, J. Polym. Sci., Polym. Chem. Ed., 15: 2793 (1977).

27. N. S. Allen, J. F. McKellar, and G. O. Phillips, J. Polym. Sci., Polym. Chem. Ed., 12: 1233 (1974).

28. N. S. Allen, J. F. McKellar, and J. F. McKellar, J. Polym. Sci., Polym. Chem. Ed., 12: 2623 (1974).

29. N. S. Allen and M. J. Harrison, Eur. Polym. J., 21: 517 (1985).

30. S. W. Beavan and D. Phillips, J. Photochem., 3: 349 (1974).

31. S. W. Beavan and D. Phillips, Rubb. Chem. Technical., 48: 692 (1975).

32. N. S. Allen and J. F. McKellar, Polym. Deg. Stab., 1: 47 (1979).

33. O. Nishimura and Z. Osawa, Polym. Photochem., 1: 191 (1981).

34. E. D. Owen and R. L. Read, Eur. Polym. J., 15: 41 (1979).

35. N. S. Allen, J. Wooler, and K. O. Fatinikun, Polym. Deg. Stab., 13: 277 (1985).

36. J. F. Rabek, R. Ranby, and T. A. Skowronski, Macromolecules, 18: 1810 (1985).

37. A. Charlesby and R. H. Partridge, Proc. Roy. Soc., A283: 312 (1965).

38. A. Charlesby and R. H. Partridge, Proc. Roy. Soc., A283: 329 (1965).

39. I. Boustead and A. Charlesby, Eur. Polym. J., 3: 459 (1967).

40. D. J. Carlsson and D. M. Wiles, J. Polym. Sci., Polym. Letta. Ed., 11: 759 (1973).

41. Z. Osawa, H. Kuroda, and Y. Kobayashi, J. Appl. Polym. Sci., 29: 2843 (1984).

42. Z. Osawa and H. Kuroda, J. Polym. Sci., Polym. Letts. Ed., 20: 577 (1982).

43. A. Charlesby and G. P. Owen, Int. J. Rad. Phys. Chem., 8: 343 (1976).

44. H. D. Scharf, C. D. Dieris, and H. Leismann, Angew Makromol. Chem., 79: 193 (1979).

45. S. W. Beavan, P. A. Hackett, and D. Phillips, Eur. Polym. J., 10: 925 (1974).

46. N. S. Allen and J. F. McKellar, J. Appl. Polym. Sci., 20: 1441 (1976).

47. Z. Osawa and K. Nagashima, Polym. Deg. Stab., 1: 311 (1979).

48. I. H. Leaver, Aust. J. Chem., 32: 1961 (1979).

49. I. H. Leaver, Photochem. Photobiol., 27: 439 (1978).

50. G. J. Smith and W. H. Melhuish, Text Res J., 46, 510 (1976).

51. K. P. Ghiggino, C. H. Nicholls, and M. T. Pailthorpe, J. Photochem., 4: 185 (1975).

52. M. D. Swords and D. Phillips, Chem. Phys. Letts., 43: 228 (1976).

53. K. P. Ghiggino, R. D. Wright, and D. Phillips, J. Polym. Sci., Polym. Phys. Ed., 16: 1499 (1978).

54. C. E. Hoyle, T. L. Nemzek, A. Mar, and J. E. Guillet, Marcomolecules, 11: 429 (1978).

55. N. S. Allen and J. F. McKellar, J. Appl. Polym. Sci., 22: 2085 (1978).

56. P. R. Cheung and C. W. Roberts, J. Polym. Sci., Polym. Letts. Ed., 17: 227 (1979).

57. N. S. Allen and J. F. McKellar, Makromol. Chem., 179: 523 (1978).

58. Y. Takai, J. H. Calderwood, and N. S. Allen, Makromol. Chem., Rep. Commun., 1: 17 (1980).

59. N. S. Allen, J. P. Binkley, B. J. Parsons, G. O. Phillips, and N. H. Tennant, Polym. Photochem., 2: 389 (1982).

60. N. S. Allen and J. F. McKellar, Makromol. Chem., 180: 2875 (1979).

61. D. J. Carlsson, L. H. Ghan, and D. M. Wiles, J. Polym. Sci., Polym. Chem. Ed., 16: 2353 (1978).

62. D. J. Carlsson, L. H. Ghan, and D. M. Wiles, J. Polym. Sci., Polym. Chem. Ed., 16: 2365 (1978).

63. M. T. Vala, R. Silbey, S. A. Rice, and J. Jortner, J. Chem. Phys., 41: 2146 (1964).

64. L. J. Basille, J. Chem. Phys., 36: 2204 (1962).

65. T. Nishihara and M. Kaneko, Makromol. Chem., 84: 124 (1969).

66. G. A. George, J. Appl. Polym. Sci., 18: 419 (1974).

67. G. A. George and D. K. C. Hodgemann, Eur. Polym. J., 13: 63 (1977).

68. W. Klopffer, Eur. Polym. J., 11: 203 (1975).

69. N. S. Allen, J. Homer, and J. F. McKellar, The Analyst, 101: 260 (1976).

70. N. S. Allen, J. Homer, and J. F. McKellar, Makromol. Chem., 17: 1575 (1978).

71. N. S. Allen, A. Parkinson, F. F. Loffelman, M. M. Ranhurst, and P. V. Sini, Polym. Deg. Stab., 7: 153 (1984).

72. N. S. Allen, J. F. McKellar, G. O. Phillips, and D. G. M. Wood, J. Polym. Sci., Polym. Letts. Ed., 12: 241 (1974).

73. N. S. Allen, D. J. Bullen, and J. F. McKellar, Chem. Ind. (London), 797 (1977).

74. N. S. Allen, D. J. Bullen, and J. F. McKellar, Chem. Ind. (London), 629 (1978).

75. N. S. Allen, J. F. McKellar, and D. Wilson, J. Photochem., 7: 319 (1977).

76. K. N. Pandey, P. S. Kanai, and V. B. Singh, Labder, J. Sci. Technol., 9-A: 220 (1971).

77. G. Winter and R. N. Whittem, Paint Notes, August: 252 (1949).

78. A. C. Somersall, E. Dan, and J. E. Guillet, Macromolecules, 7: 233 (1974).

79. H. Morawetz, Pure Appl. Chem., 52: 277 (1980).

80. R. Geller and C. W. Frank, Macromolecules, 16: 1448 (1983).

81. R. Geller and C. W. Frank, Polym. Prepr., Am. Chem. Soc., Div. Polym. Chem., 22: 271 (1981).

82. S. N. Semerak and C. W. Frank, Ad. Chem. Sers., 203: 757 (1983).

83. F. Mikes, H. Marawetz, J. Kasla, and K. S. Dennis, St. Vip. Sk. Chem.-Technol. Praze, (oddil) S., 7: 223 (1982).

84. F. Mikes, H. Morawetz, and K. S. Dennis, Macromolecules, 17: 60 (1984).

85. H. Morawetz, Polym. Eng. Sci., 23: 689 (1983).

86. X. B. Li, M. A. Winnik, and J. E. Guillet, Macromolecules, 16: 992 (1983).

87. K. Sienicki, Polym. Bull., 10: 470 (1983).

88. G. P. Egorov and E. G. Moisya, J. Polym. Sci., Part C, 16: 2031 (1967).

3
Chemiluminescence of Polymers at Nearly Ambient Conditions.

GRAEME A. GEORGE   Queensland Institute of Technology, Brisbane, Australia

I.   INTRODUCTION

The oxidation of hydrocarbon polymers is frequently accompanied by the emission of weak visible light—chemiluminescence [1].   Among the luminescence processes discussed in Chapter 1, chemiluminescence is of the lowest intensity with a typical quantum yield for light emission of 10^{-9} for liquid hydrocarbons and polyolefins [2,3].   The extreme sensitivity of modern light detection apparatus has meant that such low light levels may now be measured accurately and reproducibly [4,5].   It has been estimated that a rate of initiation of oxidation as low as 10^{-11} to 10^{-12} mol liter^{-1} sec^{-1} and radical concentrations of only 2×10^{10} radicals/cm^3 can be studied with ease by chemiluminescence [6].

While most studies of polymer chemiluminescence have been carried out at elevated temperatures (100°C or more), the ability of the technique to detect such low levels of free radicals has suggested to several workers that chemiluminescence could be measured from polymers under nearly ambient conditions [5,7].   The potential applications include the study of the following:

The kinetics of oxidation of polymers under actual service conditions and the prediction of their useful lifetime.
The mechanism of stabilization by antioxidants and UV stabilizers.
The relaxations of homopolymers, copolymers, and blends.
Changes in network structure during chain scission and crosslinking.

The effect of external and internal stress, moisture, and other
  environmental factors on polymer oxidation

   In this ~~chapter~~ paper it is intended to discuss briefly the origins of
chemiluminescence; the problems inherent in measuring chemiluminescence at low rates of oxidation; the use of perturbation techniques
and nonstationary kinetics; and the applications to the areas highlighted above. Graphs, spectra.

II. ORIGINS OF CHEMILUMINESCENCE

The apparatus required to obtain chemiluminescence data is simple
and consists of a controlled atmosphere cell with temperature stability
to ±0.2°C and provision for a cooled photomultiplier to measure the
light emitted from the sample [4,5]. The photomultiplier output is
analyzed by pulse-counting circuitry to ensure both long-term
stability and optimum signal-to-noise ratio. Because counting statis-
tics apply, the standard deviation of the light intensity measured is
the square root of the count rate. The output from the counter is
generally coupled to a microcomputer for data analysis [5]. Earlier
studies of chemiluminescence used dc circuitry, which is satisfactory
for high signals and experiments over a short time.

   The information gained from a typical polymer chemiluminescence
experiment is shown in Figure 1. This is a plot of the intensity of
light emitted as a function of the time the sample is heated at constant temperature in an oxidizing atmosphere. The earliest work of
this type reported was by Ashby [3], Schard [8], and Schard and
Russell [9]. These authors noted that

1. Significant light emission required oxygen to be present, and
    the intensity was proportional to the partial pressure of oxygen
    in contact with the surface.
2. The integrated emission intensity was proportional to the concen-
    tration of carbonyl groups produced in the oxidation.
3. Radical scavenging antioxidants decreased the intensity of
    emission and introduced an induction period before light emission
    was observed.
4. Peroxide decomposing stabilizers also introduced an induction
    period in the chemiluminescence curve.

By analogy with studies of the chemiluminescence accompanying the
liquid phase oxidation of low molecular weight hydrocarbons, it was
concluded that the emission arose from reactions of oxygenated free
radicals, most logically alkyl peroxy radicals. A source of these
was the free-radical chain oxidation of the polymer, initiated by
thermal decomposition of trace hydroperoxides incorporated during
synthesis, processing, or environmental degradation. This was

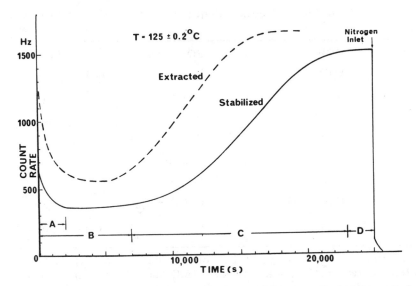

FIGURE 1   The intensity of chemiluminescence from extracted or stabilized nylon 6,6 fibers as a function of heating in oxygen after preheating to 125°C in nitrogen.   Region A shows nonstationary behavior on gas switching; region B is the induction period; region C corresponds to autoacceleration; and region D is the limiting rate region.   (From Ref. 14.)

consistent with all of the above observations and a simple reaction sequence is shown in outline in Table 1.   This does not, however, indicate the light-producing reaction; further consideration of this requires the results of the spectral analysis of the emission, quenching reactions of small molecules, and the energetics and kinetics of the free-radical reactions.

A.   Spectral Analysis

In spite of the low quantum yield, spectral distributions have been reported for chemiluminescence from polymers ranging from polyolefins [10] and polyamides [4] to epoxy resins [11] and engineering thermoplastics [12] such as poly(ether ether ketone), poly(phenylene sulfide), and polyimides.   These are generally obtained by measuring the light intensity through a series of filters and then comparing the distribution with the photoluminescence spectrum of the polymer. This is shown in Figure 2 for the emission accompanying the thermal oxidation of an amine-cured epoxy resin [11].   In common with liquid hydrocarbons, the spectral analysis is in most cases consistent with emission from carbonyl chromophores as shown in Table 2.   The low

TABLE 1 Simplified Chain Reaction Sequence for the Oxidation of a Polymer PH

Initiation	POOH	→	PO· + OH·	(1a)
	2POOH	→	PO· + PO_2· + H_2O	(1b)
Propagation	PO· + PH	→	POH + P·	(2a)
	P· + O_2	→	PO_2·	(2b)
	PO_2· + PH	→	POOH + P·	(3)
Termination	P· + P·	→	P − P	(4)
	PO_2· + P·	→	POOP	(5)
	PO_2· + PO_2·	→	POOP + O_2	(6a)
		→	P = O + O_2 + P'OH	(6b)

Note: 1a, unimolecular homolysis; 1b, bimolecular initiation; 6a, termination of two tertiary alkyl peroxy radicals; 6b, termination of at least one primary or secondary alkyl peroxy radical.

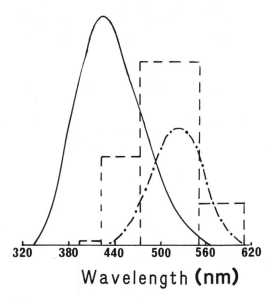

Wavelength (nm)

FIGURE 2 Luminescence spectra from an epoxy resin (tetraglycidyl-4,4'-diaminodiphenylmethane) cured with 4,4'-diaminodiphenylsulfone. (---), Chemiluminescence at 135°C in air; (——), fluorescence of uncured resin λ_{ex} 350 nm; (-·-··-), phosphorescence of cured resin λ_{ex} 430 nm.

TABLE 2 Chemiluminescence from Some Hydrocarbons and Polymers (Spectral Properties and Emitting Species)

Material	λ_{min} [a]	λ_{max}	$E_{h\nu}$ [a] (kJ/mol)	Species
Butan-2-one	450	520	270	Biacetyl [2]
Ethylbenzene	380	450	315	Acetophenone [13]
Nylon 6,6	400	500	300	Polymeric ene-ones or N-acyl amides [14]
Polypropylene	400	475	300	Polymeric carbonyl [10]
Epoxy resin (amine-cured)	390	500	307	Aromatic amide [11]
cis-1,4-Polyisoprene	390	—	307	Polymeric ene-one [15]

[a]λ_{min} is the shortest wavelength in the chemiluminescence spectrum; $E_{h\nu}$ is the corresponding energy.

energy of the emission suggests that a triplet state is populated in the reaction so that phosphorescence is observed. The excited triplet chromophore may be formed directly in the chemiluminescent reaction, by intersystem crossing from the singlet state of the carbonyl or by energy transfer from an energetic intermediate.

In all cases a lower bound to the energy provided by the chemiluminescent reaction can be determined from the emission spectrum. This is also shown in Table 2 and it can be seen that the reaction must be exoenergetic by at least 290 to 340 kJ/mol to populate the carbonyl triplet. The reactions that have been considered feasible are

1. The direct decomposition of hydroperoxides to yield an excited carbonyl and water. Vasil'ev showed that this reaction would be exoenergetic by 315 kJ/mol and proceed through a cage reaction after homolysis [16].

$$-\overset{|}{\underset{|}{C}}-OOH \longrightarrow \left[-\overset{|}{\underset{|}{C}}-O\cdot + \cdot OH \right]$$
$$\downarrow$$
$$(-\overset{|}{C}{=}O)^* + H_2O \tag{1}$$

2. Methathesis reactions of alkoxy radicals or alkyl peroxy radicals (reaction 5 of Table 1):

$$R-\underset{\underset{H}{|}}{\overset{\overset{R'}{|}}{C}}-O\cdot \; + \; P\cdot \; \longrightarrow \; (R-\overset{\overset{R'}{|}}{C}{=}O)^* \; + \; PH \qquad (2)$$

For alkoxy radical reaction this is exoenergetic by 374 kJ/mol whereas for alkyl peroxy radicals this is reduced to 323 kJ/mol.

3. Bimolecular termination of alkyl peroxy radicals (reaction 6 of Table 1). Only the reaction of pri- or sec-peroxy radicals are highly exoenergetic (460 kJ/mol) and produce an excited carbonyl group directly through a six-membered ring as an intermediate (the Russell mechanism).

$$\longrightarrow \; (-\overset{|}{C}{=}O)^* \; + \; O_2 \; + \; H \qquad (3)$$

Solution studies have favored this as the mechanism for chemiluminescence during hydrocarbon oxidation [2]. In solid polymers, however, both of the mechanisms 1 and 3 have been used to explain the data obtained from different polymers [4,7,10]. To seek a single mechanism for polymer chemiluminescence is probably over-ambitious and it may be that several light emission processes are occurring in the polymer. For example, in the liquid phase oxidation of the unsaturated hydrocarbon, linoleic acid, there is spectral and chemical evidence for emission from excited singlet oxygen [18]. Singlet oxygen emission has been well characterized in inorganic oxidation reactions [19], and the wavelengths observed and corresponding electronic transitions in the oxygen molecule are:

1269 nm: $^1\Sigma \rightarrow {}^3\Sigma$
762 nm: $^1\Delta \rightarrow {}^3\Sigma$
634 nm: $2^1\Delta \rightarrow 2^3\Sigma$
480 nm: $^1\Sigma + {}^1\Delta \rightarrow 2^3\Sigma$
381 nm: $2^1\Sigma \rightarrow 2^3\Sigma$

While the singlet oxygen can be produced directly in the Russell mechanism or by oxygen quenching of excited triplet carbonyls [3]—thus lowering the phosphorescence quantum yield—other mechanisms proposed are the direct formation from alkyl peroxy radicals [20]:

$$PO_2\cdot + PO_2\cdot \longrightarrow P\text{-}P + 2{}^1O_2{}^* \tag{4}$$

and from hydroperoxides [21]:

$$POOH \longrightarrow PH + {}^1O_2{}^* \tag{5}$$

Because of the low intensity of emission and thus broad spectral band pass in most studies of polymer chemiluminescence, the characteristic sharp emission bands from the ${}^1\Sigma$ and ${}^1\Delta$ states cannot be resolved. In addition, most photomultipliers would not detect the bands at 762 and 1269 nm. No direct evidence to support the production of singlet oxygen by either of the mechanisms indicated by reactions (4) or (5) above has been produced.

B. Kinetic Evidence

From the earliest studies of polymer chemiluminescence, it has generally been found that the intensity of emission is dependent on the extent of oxidation of the polymer. Since the concentration of polymer hydroperoxides has been shown to increase during thermal oxidation and, in the mechanism of oxidation Table 1, hydroperoxides are the initiating species as well as the major product of the propagation reaction, then Reich and Stivala [22] proposed that reaction (1) must be the chemiluminescent reaction. Other workers [10,17] have also shown that for polypropylene the experimental curves for chemiluminescence intensity as a function of time can be analyzed by a kinetic scheme based on intensity being proportional to concentration of hydroperoxide and have thus also favored reaction (1) over the bimolecular reaction (3). Against this mechanism, Vasil'ev [16] presented quenching data which support the bimolecular process. He also made the more important point that kinetic experiments which measure steady-state emission intensities cannot be used to distinguish between mechanisms (1) and (3), since in the stationary state of any linear free-radical chain reaction, the rates of initiation (reaction 1 of Table 1) and termination (at high partial pressures of oxygen, reaction 6 of Table 1) must be equal. Thus

$$I = \phi k_1 \,[POOH] = \phi k_6 \,[PO_2\cdot]^2 \tag{6}$$

where
　　I = emission intensity
　　ϕ = chemiluminescence quantum yield

Thus the isothermal chemiluminescence curve in Figure 1 shows equally well that there is an increase in the alkyl peroxy radical concentration as oxidation proceeds and also an increase in the rate of initiation due to the increased hydroperoxide concentration. As will be discussed later, nonstationary state kinetic experiments may help determine the likely mechanism.

In the chemically complex system of an autooxidizing polymer, the mechanism of light generation may change as oxidation proceeds. A simple scheme such as that in Table 1 is only a first approximation. The products of the oxidation reaction will be more readily oxidized than the substrate itself and free-radical induced decomposition of hydroperoxides may be a favored initiation mechanism.

C. Radical Reactions in Specific Polymers

While kinetic evidence alone cannot indicate the mechanism and energetic arguments favor the bimolecular reaction (3), there are obvious difficulties in applying the Russell mechanism to all polymers. For example, in polypropylene the predominant chain-carrying radical in the autoxidation of the polymer is the tertiary peroxy radical [23]:

$$
\begin{array}{c}
CH_3 \\
| \\
-C-CH_2- \\
| \\
O \\
O \\
\cdot
\end{array}
$$

From studies of model compounds, the termination rate constant of these alkyl peroxy redicals is only 10^3 to 10^4 M^{-1} sec^{-1} compared to 10^6 to 10^8 M^{-1} sec^{-1} for secondary and primary alkyl peroxy radicals [24].   Also as shown in Table 1, the products of the termination reaction are a peroxide and oxygen, and it is difficult to envisage a single-step termination reaction that can lead to chemiluminescence from a carbonyl oxidation product as indicated by the results of Table 2.   It is noted that singlet oxygen will be produced in the termination reaction which may then emit but this cannot explain the observed chemiluminescence spectrum.   However, Mayo [25] showed that termination reactions are also accompanied by the production of alkoxy radicals which will cleave to produce, ultimately, primary and secondary alkyl peroxy radicals.   These will terminate with tert-peroxy radicals via the Russell mechanism, so producing chemiluminescence from a polymeric carbonyl group.   This is consistent with results from the photooxidation of polypropylene in which the following mechanism is proposed [26]:

$$\sim CH_2 - \overset{\overset{\displaystyle H}{\underset{\displaystyle |}{\overset{\displaystyle O}{\underset{\displaystyle |}{O}}}}}{\underset{\underset{\displaystyle CH_3}{|}}{C}} - CH_2 \sim \quad \longrightarrow \quad \sim CH_2 - \overset{\overset{\displaystyle \cdot}{\overset{\displaystyle O}{\underset{\displaystyle |}{|}}}}{\underset{\underset{\displaystyle CH_3}{|}}{C}} - CH_2 \sim \quad + \quad \cdot OH$$

(7)

$$\downarrow$$

$$\sim CH_2 - \overset{\overset{\displaystyle O}{\|}}{\underset{\underset{\displaystyle CH_3}{|}}{C}} \quad + \quad \cdot CH_2 \sim$$

$$\downarrow O_2$$

$$.OO \, CH_2 \sim$$

In the solid state this β-scission reaction of alkoxy radicals may be favored over the abstraction of a hydrogen atom from the polymer. Thus, chemiluminescence from the bimolecular termination of the primary radicals can occur as a direct result of decomposition of the hydroperoxide.

Several studies of polymer chemiluminescence have been carried out in nonoxidizing atmospheres and emission is still observed. In a study of preoxidized polypropylene in nitrogen at temperatures above 130°C, the appearance of a chemiluminescence signal which decayed away over a period of 7 to 20 min (depending on temperatre) was attributed to a mechanism involving chemisorbed oxygen present as the superoxide anion [27]. It is not necessary, however, to invoke a new mechanism if oxygen is physically absorbed in the polymer since primary alkyl peroxy radicals may be formed via a reaction such as (7). In the strict absence of physically or chemically absorbed oxygen, it is still possible to develop a mechanism for chemiluminescence from polypropylene involving peroxy radical termination. It has been well known that polypropylene forms sequences of hydroperoxides at the tertiary carbon atoms along the polymer backbone (a consequence of intramolecular propagation during oxidation) [25,28]. Thus in the preoxidized polypropylene samples from which chemiluminescence is observed, the bimolecular decomposition of hydroperoxides will be favored:

$$2POOH \longrightarrow PO_2\cdot + PO\cdot + H_2O \qquad\qquad (8)$$

Analysis of the decay of chemiluminescence from polypropylene hydroperoxide under nitrogen at 110°C showed that second order kinetics was obeyed [7] and the rate constant obtained agreed with that obtained by direct chemical analysis of the hydroperoxide concentration [28]. Thus by combining Equation (8) and (7) and noting that from Table 1 (reaction 6) the termination of two tert-alkyl peroxy radicals produces molecular oxygen, the following reaction sequence is obtained for emission of chemiluminescence:

$$2\ POOH \quad \longrightarrow \quad PO_2 \cdot \quad + \quad PO \cdot \quad + \quad H_2O$$

$$PO_2 \cdot \downarrow \qquad\qquad \downarrow \ \beta\ scission$$

$$POOP + O_2 \qquad\qquad \cdot CH_2 \backsim + \quad P'C = O \tag{9}$$

$$\cdot OOCH_2 \backsim$$

$$PO_2 \cdot$$
$$\downarrow$$
$$(P\ C=O)^* \ + \ O_2 \ + \ POH$$

where for polypropylene

$$P \ = \ \backsim C \!\!-\!\! CH_2 \backsim$$
$$\mid$$
$$CH_3$$

However, as discussed in Chapter 4, first order kinetics have been observed for the decay of chemiluminescence from autooxidized polypropylene in an inert atmosphere. This would imply that isolated hydroperoxides may form under certain conditions. In the strict absence of oxygen, none of the reactions (7), (8), or (9) can explain the chemiluminescence after hydroperoxide homolysis.

Weak chemiluminescence decay from nylon 6,6 has similarly been reported when the melt-extruded film was heated under nitrogen at 100°C [29]. In a careful study of the emission it was found to disappear if the sample was treated with sulfur dioxide (a hydroperoxide decomposer); was still observed even if the sample was evacuated for lengthy periods prior to heating and the decay followed second order kinetics [29]. These results support a mechanism similar to (9) for polypropylene except that the alkyl peroxy radical in nylon 6,6 is secondary, so direct termination by the Russell mechanism of those peroxy radicals formed by the bimolecular decomposition of hydroperoxides [Equation (8)] may occur.

From a consideration of the free-radical reactions occurring during oxidation it is thus possible to rationalize the observation of many chemiluminescence experiments as resulting from termination reactions of alkyl peroxy radicals. Nonstationary chemiluminescence experiments provide further mechanistic as well as kinetic information to support this.

III. CHEMILUMINESCENCE UNDER NONSTATIONARY CONDITIONS

Apart from the initial equilibration region, the chemiluminescence curve in Figure 1 is typical of that obtained under stationary conditions. The sample is heated in a fixed atmosphere of air or oxygen at a constant temperature in the absence of light. While significant emission is observed at elevated temperatures, the intensity becomes vanishingly small in many polymers as room temperature is approached. This can be seen from a typical plot of temperature dependence of emission intensity in Figure 3 for nylon 6,6 (one of the more intense chemiluminescence emitters). In using such steady-state data for the estimation of the oxidative stability of a polymer at nearly

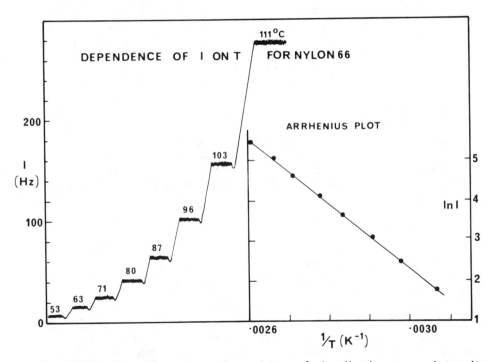

FIGURE 3 The temperature dependence of chemiluminescence intensity from nylon 6,6 in the induction period. The raw data are shown on the left and yield a linear Arrhenius plot. (From Ref. 4.)

ambient conditions, the usual procedure has been to measure the
Arrhenius parameters at elevated temperatures and then extrapolate
to room temperature on the assumption that there has been no change
in the mechanism. In several polymers, particularly those containing
stabilizers, there is evidence for a change in mechanism at lower
temperatures. A further problem in stabilized polymers is that the
induction period before significant oxidation occurs (and thus emission is observed) increases with the same Arrhenius relationship.
Thus a stationary state chemiluminescence experiment at temperatures
approaching ambient involves measuring small changes in low emission
intensities over long time periods.

An alternative approach is to perturb the stationary state of the
oxidative chain reaction by transiently increasing the concentration
of free radicals in the polymer. This will generally result in a
sudden increase in the chemiluminescence intensity followed by a
decay as the system returns to the steady state. This decay curve
may be analyzed to provide kinetic parameters for the oxidation
reaction. This has been widely applied to the liquid phase oxidation of hydrocarbons to determine the absolute value of the termination rate constant (k_6) and the rate constant for radical scavenging
by antioxidants [6]. The similarity of the chemiluminescence decay
curves obtained from solid polymers under nonstationary conditions
has led to the same kinetic scheme being applied [14]. In the following sections the kinetic relationships for chemiluminescence generated
after the application of different perturbations will be developed.
In solution oxidation studies, the nonstationary radical concentration
can be achieved most readily by the rapid addition of a free-radical
initiator. Alternative methods more appropriate for a solid polymer
are UV irradiation, a temperature pulse, gas switching, or applying
mechanical stress. In the following treatment, the light-emitting
process is initially assumed to be reaction 6 of Table 1 and the
resulting kinetic equations are then tested by comparison with experimental data.

A. Gas Switching

If a polymer is heated in an inert atmosphere in the chemiluminescence
apparatus and then oxygen or air is admitted, there is a rapid
increase in emission intensity. This was seen as region A of Figure
1 and typical curves for solid polymers are shown in Figure 4.
Polypropylene shows an initial increase followed by a slow growth to
a steady intensity while nylon, ABS (acrylonitrile-butadiene-styrene
terpolymer), and HDPE (high density polyethylene) show a burst of
emission followed by a decay to a steady emission intensity. These
diverse results may be rationalized by considering the oxidation
scheme of Table 1 and the magnitude of the rate constants.

In the absence of oxygen, the end result of the initiation reaction
1 is an alkyl radical P·. It has been found that in many polymers

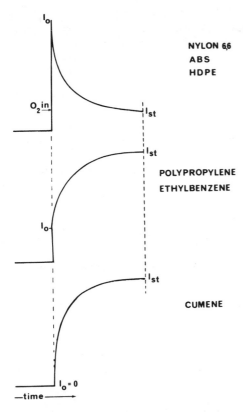

FIGURE 4 Nonstationary chemiluminescence behavior following changing the gas above the preheated polymer or liquid hydrocarbon from nitrogen to oxygen. I_0 is the instantaneous intensity after switching and I_{st} is the stedy-state value typical of that temperature. (Adapted from Ref. 4.)

the alkyl radical is long-lived and a significant concentration may therefore result [30]. The concentration of alkyl radicals will depend on the rate of the termination reaction 4 (and other alkyl radical reactions such as disproportionation not shown in Table 1), compared to the rate of initiation. In the limit of a steady-state concentration of radicals $[P\cdot]_{st}$ being achieved, the rate of initiation r_i will equal the rate of termination, i.e.,

$$[P\cdot]_{st} = \left(\frac{r_i}{k_4}\right)^{1/2}$$
(10)

If oxygen or air is now rapidly admitted to the sample, the alkyl radicals are scavenged to form alkyl peroxy radicals. At this instant, time zero in the experiment of Figure 4, the radical concentration is given by

$$[PO_2\cdot]_0 = [P\cdot]_{st} = \left(\frac{r_i}{k_4}\right)^{1/2} \tag{11}$$

The alkyl peroxy radical concentration $[PO_2\cdot]_0$ will not be the steady-state value corresponding to the temperature of the experiment, so there will be either a growth or a decay of this concentration to the steady-state value given by

$$[PO_2\cdot]_{st} = \left(\frac{r_i}{k_6}\right)^{1/2} \tag{12}$$

The kinetics of this process will be given by the solution to the equation:

$$\frac{d[PO_2\cdot]}{dt} = r_i - k_6 [PO_2\cdot]^2 \tag{13}$$

which is [31]:

$$\ln\left\{\frac{[PO_2\cdot]_t + [PO_2\cdot]_{st}}{[PO_2\cdot]_t - [PO_2\cdot]_{st}}\right\} = 2(k_6 r_i)^{1/2}t$$
$$- \ln\left\{\frac{[PO_2\cdot]_0 - [PO_2\cdot]_{st}}{[PO_2\cdot]_0 + [PO_2\cdot]_{st}}\right\} \tag{14}$$

This can be expressed in terms of the chemiluminescence intensities by Equation (6) to give

$$\ln\left[\frac{I_t^{1/2} + I_{st}^{1/2}}{I_t^{1/2} - I_{st}^{1/2}}\right] = 2(k_6 r_i)^{1/2}t + \ln\left[\frac{I_0^{1/2} + I_{st}^{1/2}}{I_0^{1/2} - I_{st}^{1/2}}\right] \tag{15}$$

Thus if $k_4 > k_6$ there will be a growth of emission to a steady value (as in polypropylene). If $k_6 > k_4$ there will be a burst of emission followed by a decay to the steady-state value, I_{st} (as in nylon 6,6). Only if $k_4 = k_6$ will there be no change in emission intensity on gas switching and the technique cannot then be used. It should be noted that r_i must be known to obtain an absolute value of the rate constant k_6 [and thus k_4' since $(k_4/k_6) = (I_{st}/I_0)$].

While Equation (15) provides a good fit to the data for gas-switching experiments during the oxidation of liquid hydrocarbons [32], the situation is much more complex in solid polymers. Figure 5a shows the chemiluminescence decay curve from nylon 6,6 fibers measured just below T_g, at 100°C, in the induction period (region A of Figure 1). The data do not fit Equation (15) over the entire time of the experiment. Such deviations from linearity could arise if a radical scavenging stabilizer is present since then Equation (13) becomes

$$- \frac{d[PO_2 \cdot]}{dt} = r_i - k_6 [PO_2 \cdot]^2 - k_7 [PO_2 \cdot] [InH] \qquad (16)$$

where InH is the inhibitor with rate constant k_7 for the peroxy radical scavenging reaction. The solution to Equation (16) will no longer be in the form of Equation (15). Extraction of the polymer to remove the stabilizer resulted in a change of the slope of the plot (Figure 5a) but it was still nonlinear. When the gas-switching experiment was carried out on the polymer at temperature above T_g and at higher extents of oxidation where all free-radical scavenging antioxidants have been consumed, a linear plot was obtained (Figure 5b). This may indicate that not all of the stabilizer was removed by extraction, but an alternative explanation involves the different mobility of radicals in the solid state. One explanation of the non-linearity involves geminate recombination of peroxy radicals formed in the initiation cage prior to diffusion apart by normal chain mobility. The chemiluminescence decay will thus be the sum of a rapid and a slow process. Billingham [33] considered that the mobility of the peroxy radicals should also include a component due to the propagation reaction 3 of Table 1. This explains both the nonlinearity of the decay curve and the much shorter lifetime of the alkyl peroxy radical compared to the alkyl radical.

The gas-switching experiments on polymers can provide important information about radical mobility in the solid polymer but the complexity of the process has so far prevented a quantitative analysis. The technique requires a significant rate of initiation and this means that the sample must generally be heated above ambient temperatures. However, it can be effectively combined with other perturbation techniques discussed below.

B. Ultraviolet Irradiation

Commercial polymers contain a range of UV-absorbing chromophores which may lead to free-radical formation. These include hydroperoxides, peroxides, carbonyl groups, polynuclear aromatics, as well

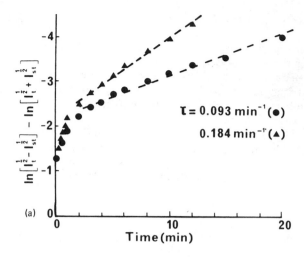

FIGURE 5a   Analysis of the decay of the chemiluminescence pulse from nylon 6,6 at 100°C in the induction period on switching from nitrogen to oxygen according to Equation (15).   ▲ stabilized, ● extracted.   (From Ref. 14.)

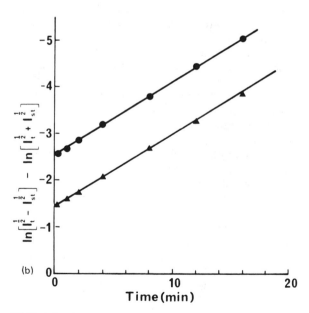

FIGURE 5b   Analysis of the decay curves from extracted nylon 6,6 at 137°C in the limiting rate region after admitting oxygen to the sample previously heated in nitrogen (●) or in air (▲).   (From Ref. 14.)

as residual catalysts and stabilizer decomposition products. From studies of the concentration of these chromophores in commercial polypropylene and the quantum yield for free-radical generation, the hydroperoxide group is considered the primary photoinitiation center [34]. Thus by UV-irradiating a polymer in the chemiluminescence apparatus in the presence of air or oxygen, it is possible to create a transiently high concentration of free radicals of the same species as would be produced by thermal decomposition of the hydroperoxide [reaction 1 of Table 1, and Equation (7)]. The rate of photoinitiation r_i' is given by

$$r_i' = \frac{f\phi' I_a}{V} \tag{17}$$

where

f = efficiency of radical escape from the initiation cage
ϕ' = photochemical quantum yield
I_a = intensity of light absorbed by the chromophore
V = volume irradiated

When the polymer is irradiated in air, a new steady-state concentration of alkyl peroxy radicals will be reached in which

$$r_i' = k_6 [PO_2 \cdot]_0^2 \tag{18}$$

If UV irradiation is now stopped, there will once again be a transiently high concentration of radicals which will decay back to the steady-state value at that temperature. If r_i is small, e.g., the polymer is at room temperature, then the return to the steady state is given by

$$\frac{d[PO_2 \cdot]}{dt} = - k_6 [PO_2 \cdot]^2 \tag{19}$$

or in integral form:

$$\frac{1}{[PO_2 \cdot]_t} - \frac{1}{[PO_2 \cdot]_0} = k_6 t \tag{20}$$

Converting to emission intensities using Equation (6) and noting that the radical concentration at time zero is given by Equation (18), the decay of chemiluminescence is described by

$$\frac{1}{I^{1/2}} - \frac{1}{I_0^{1/2}} = \left(\frac{k_6}{\phi} \right)^{1/2} t \tag{21}$$

Thus a plot of $I^{-1/2}$ against time should yield a straight line, slope $(k_6/\phi)^{1/2}$, and intercept $I_0^{-1/2}$. The radical half-life $\tau = (k_6 [PO_2]_0)^{-1}$ is given by the ratio of the intercept to the slope. Since second order kinetics are followed, it should be noted that the radical half-life depends on the initial radical concentration.

Again, as for gas switching, a simple second order kinetic equation will not be expected if stabilizer reactions are occurring or if there is a nonuniform distribution of peroxy radicals over the volme analyzed.

As this technique enables radical half-lives and rate constants to be studied at temperatures approaching ambient, it is the most generally useful nonstationary technique. Examples of the use of Equation (21) will be given in Section IV, with applications ranging from the study of stabilizer reactions in polyolefins to the measurement of the change in network structure during the crosslinking of an epoxy resin. In many cases the simple second order kinetics of Equation (21) are followed.

Quite recently, Mendenhall [65] suggested that the long-lived emission observed from many polymers after irradiation may be explained by the charge recombination of photo-ejected electrons. In a study of 98 commercial polymers, about one-third of them showed a hyperbolic luminescence decay after irradiation. This, together with the independence of the emission decay properties to the presence of moisture, oxygen, and temperature over the range 6 to 40°C, indicated the emission was not of free-radical origin. The ability to excite emission with even visible light supported a mechanism such as the trapped electron ion recombination that occurs in radiothermoluminescence [66]. As briefly indicated in Chapter 1 and discussed at length in Chapter 7, thermoluminescence is normally observed on heating a polymer sample previously irradiated at low temperatures. Emission is then observed as a glow curve in the range 77 to 300K. The process reported by Mendenhall occurs immediately after irradiation at near-ambient temperatures and most often corresponds to recombination of shallow-trapped electrons with the positive ion remaining after photo ejection from either the polymer chromophore or an impurity center [65].

It is possible that this process is occurring together with the chemiluminescence arising from peroxy-radical recombination in certain polymers. It may be readily distinguished from chemiluminescence by performing the entire irradiation and luminescence measurement with the polymer in an inert atmosphere where peroxy radical formation by scavenging of alkyl radicals cannot occur. In the studies of the polyamides discussed in Section III.D, emission in an inert atmosphere is not observed, and a mechanism based on peroxy radical recombination leading to a decay curve described by Equation (21) is seen to describe the nonstationary chemiluminescence.

The situation with other polymers is not as clear-cut and further research is required.

C. Mechanical Stress

From electron spin resonance studies of the tensile deformation of oriented polyolefin and polyamide fibers it has been shown that free radicals are formed when the fiber is strained to >60% of ultimate. This has been attributed to the progressive rupture of taut tie molecules in the amorphous region of the polymer [35]. In air, the alkyl radicals so formed will be readily scavenged by oxygen to form alkyl peroxy radicals which will, on termination, yield chemilumi- nescence. The spectral distribution of this emission is identical to thermal chemiluminescence and the net effect of the applied stress is to lower the activation energy for the oxidation of the polymer [36]. The rate of free-radical formation r_b on application of a stress σ to the polymer chain at a temperature T (K) is given by the Zhurkov equation:

$$r_b = n_c \omega_0 \exp\left[\frac{-(U_0 - \beta\sigma)}{RT}\right] \qquad (22)$$

where n_c is the number of tie molecules available for scission, ω_0 is the fundamental bond vibration frequency, U_0 is the main-chain bond strength, and β is the activation volume for bond scission [35].

Thus if a stress σ is applied to the polymer and then removed, there will be a transient increase in the chemiluminescence intensity followed by a decay to the steady state. This will be described by Equation (21) except that r_i' is replaced by r_b from Equation (22). Figure 6a shows such an experiment [36] in which medium-tenacity nylon 6,6 yarns are strained to failure in air at 40°C and the chemi- luminescence measured. From Figure 6b it is seen that the emission intensity increases exponentially with applied stress as expected by Equation (22). When the stress is removed (in this case by failure of the fibers) there is decay of the emission. In Figure 6c is shown a plot of the decay of this emission after fiber failure according to second order kinetics. The plots are linear and yield a peroxy radical half-life that is independent of temperature around 40°C. This suggest that the termination reaction of alkyl peroxy radicals occurs within the activation volume in which chain scission has taken place. It is seen that above 60°C the decay becomes complex and the half-life increases indicating the termination of radicals propagat- ing outside the initiation cage. Such studies, coupled with stress relaxation and cyclic loading experiments, can provide valuable infor- mation on the oxidation rate and thus service lifetime of polymers under load. The entire phenomenon of stress chemiluminescence is considered in a separate chapter in this book.

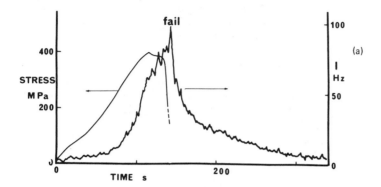

FIGURE 6a The stress-time and chemiluminescence-time curves from medium tenacity nylon 6,6 yarns strained in air at 10% min^{-1} up to failure. (From Ref. 36.)

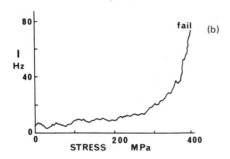

FIGURE 6b The plot of chemiluminescence intensity against fiber stress showing the exponential increase in intensity according to Equation (22). (From Ref. 36.)

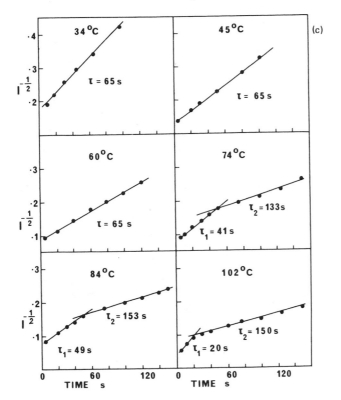

FIGURE 6c Second order kinetic analysis of the decay of chemiluminescence after fiber failure as shown in Figure 6a. The radical half-
life τ is marked for each experimental temperature. (From Ref. 36.)

D. Combined Perturbations

It is possible to probe particular features of free-radical reactions in
a polymer by combining any of the above perturbation techniques.
One such example is the combination of UV irradiation and gas switching as shown in Figure 7 for an experiment with nylon 6,6 fibers at
30°C [7]. The initial UV irradiation was carried out for 2 min in
nitrogen and then irradiation was ceased and the luminescence recorded a few seconds later. The following features of the chemiluminescence are noted:

1. Initially, the photomultiplier detects intense phosphorescence
    from the sample. (As noted in Chapter 1, the quantum yield of
    room temperature phosphorescence from polyamides is quite high.)
    This photoluminescence decays away over 60 sec until only the
    low background count is observed. No chemiluminescence is
    observed in nitrogen.

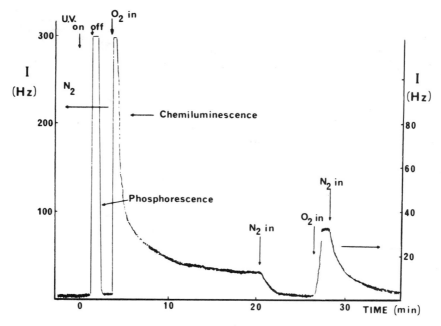

FIGURE 7   Phosphorescence and chemiluminescence from nylon 6
fibers at 30°C after UV irradiation in nitrogen for 2 min and the
atmosphere then changed as shown.   (From Ref. 7.)

2.   As soon as oxygen is admitted to the sample, there is a burst
     of chemiluminescence followed by a slow decay over tens of
     minutes.
3.   Flushing the sample with nitrogen results in a more rapid decay
     of the chemiluminescence to the background count.
4.   If oxygen then replaces nitrogen, chemiluminescence is observed
     once more.   This sequence of alternate exposure of the sample
     to oxygen and nitrogen generates the chemiluminescence pattern
     of Figure 7 for periods of hours after the original short UV
     irradiation.

It would be expected that the decay of chemiluminescence after
admission of oxygen would follow the decrease in alkyl peroxy radical
concentration and second order kinetics as described by Equation
(21) would be obeyed.   A second order plot of the decay curve is
shown in Figure 8 and it is seen that the plot is curved.   However
when the decay curve on admitting nitrogen in feature 3 above is
plotted, second order kinetics are obeyed.   It thus appears that the
production of alkyl peroxy radicals ceases only when nitrogen is
admitted and the decay curve reflects the radical lifetime (in this

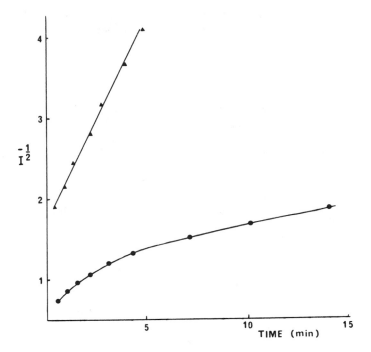

FIGURE 8   Second order analysis of the chemiluminescence decay
curve of Figure 7 according to Equation (21): •, after the initial
admission of oxygen; ▲, after nitrogen is readmitted. (From Ref.
7.)

case 3.7 min). When oxygen is present, the decay curve must
represent the sum of a slow, diffusion-controlled radical formation
reaction and the termination reaction.   These results support the
suggestion in Section III.B from thermal chemiluminescence that the
alkyl radicals in nylon 6,6 are extremely long-lived and scavenging
by oxygen is a diffusion-controlled process. Unfortunately, such a
result makes the kinetic analyses of nonstationary gas-switching
experiments difficult as the solutions such as Equation (15) assume
instantaneous scavenging of all alkyl radicals to become alkyl peroxy
radicals. An important application of the result concerns the best
technique for stabilizing the polymer and this is discussed later in
Section IV.B.

     A combined perturbation method that has not been explored is
UV irradiation combined with mechanical stress. Most of the kinetic
data for the photooxidation of polymers has been generated from
polymer powders and films in the absence of stress. However, when
they are used as fabricated articles they may be subjected to static

or cyclic loading, and will in many cases have considerable internal
stress resulting from processing. The nonstationary methods have
the potential for providing kinetic information under these combined
environmental perturbations.

E.   Mechanistic Implications

In the above nonstationary experiments, the chemiluminescence decay
data has been explained as following second order kinetics, with the
assumption being made that this results from the bimolecular termination reaction of alkyl peroxy radicals.   As discussed before in
Section II.A, an alternative mechanism used by several authors [10,
17,37] is homolysis of polymer hydroperoxides giving chemiluminescence from an excited ketone produced by the cage reaction of the
alkoxy and hydroxyl radicals [Equation (1)].   Using this mechanism,
the observation of second order kinetics could be rationalized as a
consequence of the bimolecular initiation reaction of adjacent hydroperoxides—a well-characterized reaction in polypropylene [Equation
(1)].   Indeed, second order chemiluminescence kinetics are found to
be obeyed for the decomposition of polypropylene hydroperoxide
under steady-state conditions in nitrogen [7].   However, it was
shown earlier (Section II.B) that steady-state experiments cannot
distinguish between chemiluminescence arising from initiation and
termination reactions (since in the steady state, the rates of the two
reactions must be equal).   It is therefore of interest to examine the
nonstationary data produced by different perturbation methods to
see if hydroperoxide homolysis [Equation (1)] can explain the kinetic
data.

   This mechanism would require that the perturbation produce a
transient increase in hydroperoxide concentration and the decay of
the chemiluminsecence would represent the rate of reaction 1 in
Table 1.   Since the emission results from a rapid cage reaction, the
perturbation cannot produce an increase in just the rate of decomposition of the hydroperoxide, since when the perturbation, such as
UV irradiation, is removed, the return to the steady state will be
extremely rapid.   This mechanism cannot therefore explain the decay
of chemiluminescence over time periods of minutes and even hours
shown in Figure 7.   While it would be expected that UV irradiation
of the polymer in air or stressing it to produce chain scission may
ultimately lead to an increase in the hydroperoxide concentration,
when the perturbation is removed the rate of decomposition of the
hydroperoxide must instantly become zero in those experiments
carried out at temperatures near ambient.   For example, at 25°C
the rate constant for decomposition of polypropylene hydroperoxide
is 10^{-10} sec^{-1}.   The data presented in Figures 6 and 7 cannot be
rationalized as resulting from chemiluminescence arising from a cage
reaction as required by this mechanism [Equation (1)].

It is concluded that the slow decay of chemiluminescence must reflect reactions of polymer macro radicals that have escaped the initiation cage. Of the reactions discussed in Section II.B, only the metathesis reaction of alkoxy radicals with alkyl radicals [Equation (2)] or the bimolecular termination reaction of peroxy radicals [Equation (3)] meet the energetic requirements. It is difficult to rationalize the gas-switching data of Figure 7 with a reaction scheme [Equation (2)] that is first order in alkyl radical concentration, since on admission of oxygen the alkyl radical concentration must decrease, but the chemiluminescence intensity is seen to rapidly increase. The only mechanism that is consistent with the energetic requirements, and both steady-state and nonstationary chemiluminescence spectral and kinetic data, is the bimolecular termination of alkyl peroxy radicals. This will be used in discussion of the applications of chemiluminescence.

IV. APPLICATIONS

The following applications have been chosen to illustrate the approaches that are being developed to enable the chemiluminescence technique to be used under conditions approaching ambient, although the data so far obtained may be at an elevated temperature or other conditions that favor a higher chemiluminescence intensity. While the technique can provide useful kinetic, mechanistic, and structural information, it must be supported by other sensitive methods of analysis such as micro oxygen uptake, Fourier Transform Infrared (FT/IR) spectroscopy, Electron Spectroscopy for Chemical Applications (ESCA), and thermal analysis, if its full potential is to be realized.

A. Degradation Under Ambient Environmental Conditions

Much of the understanding of the mechanism of the degradation and stabilization of polymers has come from applying the concepts embodied in the free-radical chain reaction mechanism for the thermal and photochemical oxidation of liquid hydrocarbons. Experimental studies have involved oxygen uptake or oxidation product measurement by spectroscopic methods (UV, IR, and luminescence). Typical oxygen uptake results during the photooxidation of polypropylene at a UV dose rate of 880 W/cm^2 [38] are shown in Figure 9, and it is seen that there is a low but steadily increasing rate of oxidation in the "induction period" that increases to the steady-state value. In this period, the weight average molecular weight $\overline{M}_w$ has decreased from 250,000 to around 40,000, corresponding to five chain scissions per molecule. The material is thus highly degraded and embrittled before the end of the induction period, so that if the kinetics of oxidation are to be used to predict the useful lifetime of this polymer, the data must be generated in the early stages of oxidation and not in the steady-state region. Mayo [39] similarly noted for heat aging

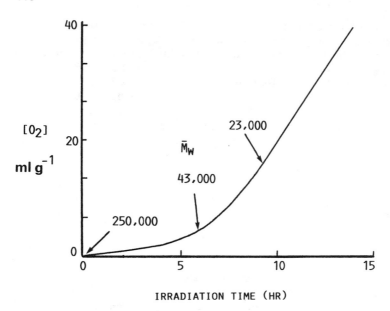

FIGURE 9 Oxygen uptake curve during the photooxidation of a 60-µm polypropylene film. The changes in $\bar{M}_W$ during irradiation are marked. (Adapted from Ref. 38.)

of polymers that the kinetics of oxidation beyond the induction period are unimportant since most polymers embrittle in this time. In addition, in the presence of stabilizers, the nature and concentration of oxidation products is altered, and in the example of fiber photooxidation no correlation was found between oxidation product buildup and embrittlement. Such results must be kept in mind when chemiluminescence is used to study the rate of oxidation and then used to predict service lifetimes.

In an early comparison of the chemiluminescence from autooxidizing polypropylene with the extent of oxidation measured by carbonyl concentration from infrared spectrophotometry, it was shown that the rate of oxidation could be measured with much greater sensitivity and at short times by means of chemiluminescence [8]. While this study was carried out at 150°C, the use of modern photon-counting techniques has enabled chemiluminescence to be detected as low as 25°C in nylon 6,6 [5] and 35°C in polypropylene [40]. In the latter study, steady-state chemiluminescence from a range of polyolefins was studied after γ irradiation and correlated with the elongation at break of the polymers when mechanically tested. In this and other studies of γ-irradiated polyolefins [41,42] it has been assumed that emission arises from bimolecular termination of alkyl peroxy radicals.

No account appears to have been taken of the thermoluminescence
from annihilation of ionic and radical species previously trapped in
the crystalline region during irradiation [43]. The lowest temperature at which steady-state chemiluminescence may be used to measure
oxidation kinetics in polyolefins may therefore be somewhat higher.
A recent study of polystyrene has demonstrated the high sensitivity
of chemiluminescence in detecting surface oxidative degradation compared with absorption spectrophotometry and gel permeation chromatography [44] but temperatures above 60°C appear to be required
for significant steady-state emission intensities.

The measurement of the activation energy for the initiation of
oxidation has frequently been determined [1,4,5,9] by steady-state
chemiluminescence from an Arrhenius plot of the emission intensity
as shown in Figure 3. Since in the steady state:

$$I = \phi r_i \tag{23}$$

where r_i is the rate of initiation, the slope of the line in Figure 3
will also depend on the temperature dependence of ϕ, the chemiluminescence quantum yield. In many cases it has been assumed that
this is negligible, since a linear Arrhenius plot is obtained even
when the temperature range encompasses a polymer relaxation such
as T_g [4]. This is not always so and, as discussed in Section IV.C,
it may be used to study relaxations. Certainly, close agreement
has been reported between the chemiluminescence method and a
chemical method for measuring the activation energy for initiation of
oxidation in solution [32]. A discontinuity in the Arrhenius plot
has been used to detect changes in the initiation mechanism with
temperature as ambient conditions are approached [1] and has also
been noted in a polyolefin containing a peroxide-decomposing antioxidant [8]. While a value for the activation energy for the initiation of oxidation is of use in determining the mechanism of initiation
and any discontinuities in the Arrhenius plot may indicate a change
in mechanism, the absolute value of the initiation rate is required
for determining the kinetics of oxidation.

In principle, the absolute value of the initiation rate may be
determined by methods used in solution oxidation kinetics, such as
the method of inhibitors [6]. By measuring the induction time for
the oxidation reaction as a function of concentration of an alkyl
peroxy radical-scavenging antioxidant (such as a hindered phenol),
the rate of inhibition can be measured. At high initial concentrations
of inhibitor this will be equal to the rate of initiation. Since the
induction period of a stabilized polymer can be readily measured by
chemiluminescence [45], the rate of initiation can in principle be
determined. Attempts to apply this technique in solid polymers
appear to be limited by the presence of zones of unequal radical

initiating and scavenging rates, even in amorphous polymers such as polystyrene [46].

In most polymers, detailed information about initiating species and their reactions is lacking because of the low concentration of the impurity centers which initiate the oxidation reaction. As ambient temperatures are approached, this becomes increasingly difficult as the rate of radical formation is vanishingly small. An alternative approach is to incorporate higher concentrations of the suspected initiating species—such as the hydroperoxidized polymer—in the material studied, and then observe the rate of radical formation and the effect of stabilizers by following the chemiluminescence intensity. In this case, significant changes in the concentration of the initiating species in the time frame of the chemiluminescence experiment will only occur at elevated temperatures. However, once the mechanism of initiation and the rate constant have been determined, the experiment may be extended to lower temperature.

If the initiating species, generally a hydroperoxide, POOH, decomposes by either a unimolecular or bimolecular process, the chemiluminescence intensity at any time t is given by

Unimolecular: $I_t = I_0 \exp(-k_1 t)$ (24)

Bimolecular: $\left(\dfrac{I_0}{I_t}\right)^{1/2} - 1 = k_2 [POOH]_0 t$ (25)

By plotting the chemiluminescence data according to Equations (24) and (25), both the mechanism and the rate constant at any particular temperature may be determined. When polypropylene containing 0.17 mol/kg of its hydroperoxide was studied by chemiluminescence in this way, it was found that bimolecular initiation occurred and the rate constant k_2 determined from Equation (25) was close to that by conventional chemical analysis [36]. It was also shown that a hindered amine stabilizer (Tinuvin 770) and its oxidation product (a nitroxyl radical) were able to decompose the hydroperoxide at a rate 10 times higher than normal thermal decomposition. This chemiluminescence data confirmed results obtained by ESR and FT/IR spectroscopy [47] and could in principle be extended to lower temperatures.

A similar approach has been used by Mendenhall et al. [48] in a study of the oxidation of 1,4-polyisoprene, 1,2- and 1,4-polybutadiene, and trans-polypentenamer at temperatures near ambient [48]. In order to generate sufficient hydroperoxides for examination at 28°C, the polymers were either autoxidized at elevated temperatures or reacted with singlet oxygen. Photosensitized oxidation of the polymer films containing added singlet oxygen sensitizers showed nonstationary chemiluminescence behavior. The decay curve fitted second order kinetics as predicted by Equation (21) but no analysis was presented [49].

Among the nonstationary techniques applied to studies of degradation reactions near ambient, UV irradiation and gas switching (or both combined) as described earlier in Section III have been explored. Nylon 6,6 as both film and fibers has been most closely studied [4, 5,7,14,29,33], and among the results of the chemiluminescence studies are the following:

1. During thermal oxidation in air, significant concentrations of long-lived alkyl radicals are formed.
2. The termination rate constant of alkyl peroxy radicals is eight times that of alkyl radicals.
3. The rate of termination of alkyl peroxy radicals includes a component due to movement through the polymer by chain propagation [29].
4. Because of the observation of alkyl radical reactions in air, effective stabilization of the polymer requires an alkyl radical-scavenging antioxidant as well as alykl peroxy radical scavengers such as hindered phenols.

B.   The Mechanism of UV Degradation and Stabilization

The UV perturbation technique of Section III.B provides a direct method for studying both photoinitiation and the reactions of stabilizers with the resulting radicals. This may be demonstrated by the following data for the UV irradiation of hydroperoxidized polypropylene (containing 0.17 mol/kg POOH) at 30°C for different times. The decay of chemiluminescence after ceasing irradiation is analyzed according to second order kinetics [Equation (21)] and it can be seen from Figure 10a that a linear plot is obtained. Also shown in this figure are the analyses of the chemiluminescence decay after UV irradiation for 90 sec of the hydroperoxidized polypropylene containing 0.5 wt% of the hindered amine light stabilizer (HALS) Tinuvin 770 and its dinitroxyl radical (HALSNO·). It is seen that only the nitroxyl radical shortens the lifetime of the chemiluminescence, a clear indication that it is scavenging the radical products of photodissociation of the hydroperoxide. This is in contrast to the results described in the previous section in which both the amine and its nitroxyl radical were able to increase the rate of chemiluminescence decay. This suggests that the amine increases the rate of decomposition of the hydroperoxide rather than scavenging free radicals from the thermooxidative degradation initiated by polypropylene hydroperoxide at 110°C.

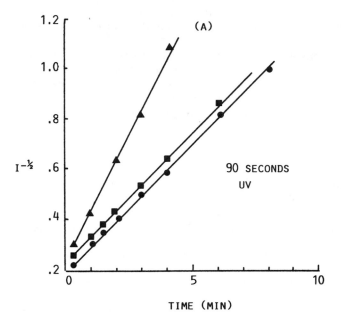

FIGURE 10a   Second order analysis of the chemiluminescence decay
after UV irradiation for 90 sec of •, hydroperoxidized polypropylene
PPOOH (0.17 mol/kg); ▪, PPOOH + 0.5 wt% HALS (hindered amine
light stabilizer); ▲, PPOOH + 0.5 wt% HALS NO· (nitroxyl radical
of HALS).

In Figure 10b is shown the analysis of the decay after 600 sec
UV irradiation in air at 30°C.   Now there is evidence that the sample
containing amine is showing radical-scavenging activity although it
is much less effective than the nitroxyl radical.   It is known that
hindered amines are oxidized to the nitroxyl radical in the course
of photooxidation and this is the active radical scavenger [50,51].
The principal route for the formation of the nitroxyl is considered
to be reaction with the alkyl peroxy radical during polymer oxidation:

$$\text{N--H} \quad + \quad \text{PO}_2\cdot \quad \rightarrow \quad \text{N--O}\cdot \quad + \quad \text{POH} \qquad (26)$$

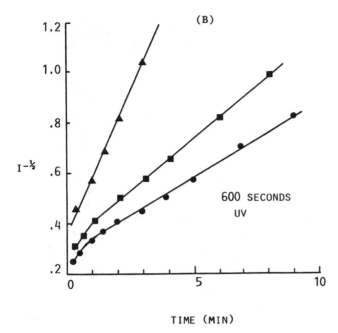

FIGURE 10b   Repeat of the experiment in Figure 10a but with UV irradiation for 600 sec.

Reaction with singlet oxygen or direct reaction with hydroperoxide may also generate the nitroxyl radical, but in the presence of UV radiation Equation 26 adequately explains the observation of radical scavenging by the sample containing only hindered amine.   Studies of the chemiluminescence decay parameters from hydroperoxidized polypropylene without stabilizers and irradiated for different times show an increase in the radical lifetime with time of irradiation. This may indicate a nonuniform distribution of free radicals in the polymer which complicates the kinetic analysis.

Nylon 6,6 shows strong UV-stimulated chemiluminescence at ambient temperatures as shown in Figure 7.   The occurrence of long-lived alkyl radicals following UV irradiation as well as during thermal oxidation means that for effective photoprotection a stabilizer which scavenges alkyl radicals as well as alkyl peroxy radicals is required.   The hindered amines have been shown to function in this way [50,51] and it would be of interest to study their stabilization reactions in nylon by chemiluminescence in the same way as reported above for polypropylene.

C. Polymer Relaxations

Luminescence techniques have been widely used for studying relaxation and transport properties in polymers [52]. In many cases the
luminescence quantum yield is sensitive to the polymer free volume,
resulting in a change in emission intensity and lifetime on passing
through a transition as the temperature is increased. This should
also apply with chemiluminescence since both the quantum yield ϕ
and termination rate constant k_6 of Equation (6) should depend on
the free volume and local mobility of the polymer chain containing
alkyl peroxy macro radicals. It would be expected that if the temperature range encompassed T_g (the glass-rubber transition), then
a sharp discontinuity in the temperature dependence of chemiluminescence intensity would occur at this point. This has been observed
from crosslinking epoxy resins [52,53] in which, as T_g was changed
by altering the ratio of the resin and hardener and the extent of
postcure, there were systematic changes in the chemiluminescence-
temperature plot. This provided a simple noncontact method for
measuring T_g of epoxy resin coatings [53]. In such studies it must
be recognized that the rate of free-radical formation will also be
increasing with temperature, depending on the activation energy for
the oxidation reaction. In some polymers, such as nylon 6,6, this
apparently masks the transition since, as shown in Figure 3, a linear
dependence of log I on 1/T is observed over a temperature range
that encompasses T_g.

In studies of the temperature dependence of chemiluminescence
from compatible blends of polystyrene and poly(vinylmethylether),
it was found that sharp changes in emission intensity occurred at
progressively higher temperature as the polystyrene content of the
blend was increased [54]. This temperature appeared to be higher
than T_g of the blend and corresponded to phase separation of the
components. This was found to be more sensitive than cloud point
measurements.

D. Network Structure: Crosslinking and
     Scission

In the previous section it was noted that chemiluminescence was
dependent on free volume and could be used to determine T_g of a
crosslinked network—an epoxy resin. It has also been reported
[53,55] that there are systematic changes in both the steady-state
and nonstationary chemiluminescence properties of the epoxy resin
as the newtork is formed during the crosslinking reaction of the
glycidyl groups of the epoxy resin with the amine groups of the
hardener. The cure of the resin takes place in the steps of gelation
followed by vitrification which may be studied by thermal analysis,
dielectric relaxation, and dynamic mechanical analysis techniques to
pinpoint the gel point, at which a well-developed system of crosslinks

is formed. In the industrial process of fabrication of fiber-reinforced composite materials the crosslinking reaction must be monitored continuously to ensure that wet-out and compaction occur prior to gelation and so produce a laminate that is free of voids [55].

It has been shown by comparison with thermal analysis that chemiluminescence can be used to monitor the critical gelation phase of network formation [11]. The reasons for this are discussed in the text that follows:

The gelation of an epoxy resin results in an increase in resin viscosity from around 1 to 10^{-3} poise. The epoxy resin and hardener undergo trace oxidation at the cure temperature to produce alkyl peroxy radicals [11] which will diffuse together and terminate to produce chemiluminescence. From Table 2 and the spectral analysis shown in Figure 2, in amine-cured epoxy resins the emitting species is the excited triplet state of an amide. The activation energy for the termination reaction is negligible so the reaction becomes diffusion-controlled. This means that k_6 in reaction (6) becomes the diffusion-controlled value k_d as given by the Debye equation:

$$k_d = \frac{8RT}{3\eta} \tag{27}$$

where η is the viscosity. Thus as gelation takes place k_d will decrease by a factor of 1000, with a resultant decrease in chemiluminescence intensity. Such an experiment is shown in Figure 11 for an epoxy resin (MY720-DDS)[*] containing different concentrations of hardener and thus having different gel times. The sharp decrease in emission intensity in both samples was found to agree exactly with the gel point as measured by thermal analysis [11,55].

Systematic changes are also observed in the nonstationary chemiluminescence parameters during formation of a crosslinking network. The most readily applied perturbation at all temperatures is UV irradiation, so the kinetic equation describing the decay of chemiluminescence following the cessation of irradiation in air is Equation (21). This second order plot yields a straight line with the slope of the line being $(k_d/\phi)^{1/2}$ and the ratio of the intercept to the slope giving the radical lifetime τ. In Figure 12 are shown the changes to these parameters during cure of the epoxy resin MY720 with two different concentrations of the hardener DDS. It is seen that the parameter $(k_d/\phi)^{1/2}$ decreases as the viscosity increases as expected from Equation (27), passes through a minimum at the gel point and then increases again. This suggests that up to gelation the parameter

[*]The epoxy resin Ciba-Geigy MY720 consists mostly of tetraglycidyl-4,4'-diaminodiphenylmethane and is cured with Ciba-Geigy Eporal DDS hardener which is 4,4'-diaminodiphenylsulfone.

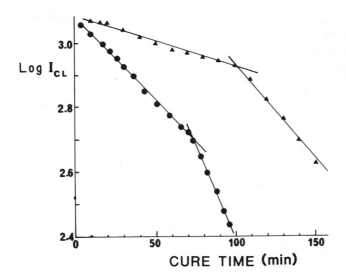

FIGURE 11   The change in steady-state chemiluminescence intensity
during the reaction of the epoxy resin MY720 with (▲) 27% and (•)
37% of diaminodiphenylsulfone (DDS).   The change in slope corresponds to the gelation of the resin.   (From Ref. 55.)

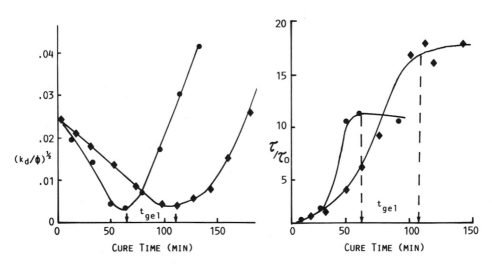

FIGURE 12   The change in the nonstationary chemiluminescence
parameters during the reaction of the epoxy resin MY720 with (♦)
27% and (•) 37% of diaminodiphenylsulfone (DDS).   The gel time
(t_{gel}) from thermal analysis is indicated.   (From Ref. 55.)

is dominated by the decrease in k_d as viscosity increases, but once a crosslinked network is formed there is little further change in the termination rate constant. This is supported by the curve for the radical half-life τ or $(k_d[PO_2.])^{-1}$, which is seen to level off at the gel point. The increase in $(k_d/\phi)^{1/2}$ beyond gelation suggests a decrease in ϕ, the chemiluminescence quantum yield, or perhaps a sharp decrease in oxygen solubility means fewer peroxy radicals are formed. It would be expected that as the resin vitrifies the radical mobility would approach zero. The chemiluminescence decay curves at temperatures below T_g do not follow simple kinetics and there appear to be short- and long-lived radicals generated on UV irradiation. This would be consistent with the view of the rather inhomogeneous nature of cured epoxy resin systems [56]. Chemiluminescence would appear to offer a suitable technique for monitoring network formation in fiber-reinforced composites under industrial fabrication conditions by using fiberoptics to collect the emitted light. It is of interest to note that chemiluminescence during network formation can also be seen in the presence of high loadings of carbon fibers in the composite materials, although the intensity is greatly reduced, and gelation may be readily followed by the above technique [55].

The nonstationary chemiluminescence technique may also be used to follow the reverse process of network breakdown by chain scission. In this case there should be a decrease in the radical half-life since the chain scission will increase the local mobility and thus k_d. The effect of time of accelerated UV exposure (>290 nm) on two different crosslinked epoxy resins is shown in Figure 13. The MY720-DDS epoxy resin is cured above 135°C and has a high crosslink density (because MY720 is a tetrafunctional epoxide, as described before) and there is an immediate decrease in the radical half-life. In contrast, the room temperature-cured system is based on Ciba-Geigy GY250, which is a diglycidyl ether of bisphenol A. This is seen, on UV exposure, to show an initial increase in the radical lifetime. This could indicate that further curing of the resin is taking place at the temperature of the accelerated weathering device (40°C). However, after this initial increase there is a more rapid decrease in half-life than from the tetrafunctional MY720-DDS. This reflects the lower crosslink density resulting from curing of the difunctional epoxide, so that each chain scission event leads to a greater increase in free volume and thus a decrease in lifetime. The UV degradation properties were also found to depend on the amount of amine hardener used. This may simply reflect a change in the initial crosslink density, although the initiation reaction of photodegradation involves the photochemistry of the amine group [57].

The decrease in the crosslink density on UV exposure as measured by the chemiluminescence parameters is confined to a thin surface layer since when $\sim$10 µm of the surface of the GY250-HY837 resin

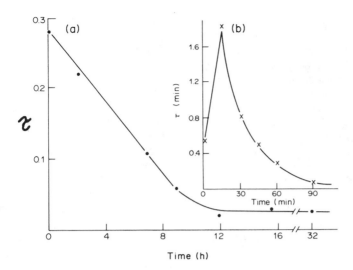

Time (h)

FIGURE 13   The change in radical half-life τ of cured epoxy resins
with time of exposure to UV in an accelerated weathering device:
•, MY720 cured with 17% DDS at 135°C; x, GY250 cured with 26%
HY837 at room temperature.   (From Ref. 53.)

was removed after 90 min irradiation (Figure 13), the radical half-
life returned to a value close to that from the original material.
This result agrees with the reported surface sensitivity of chemiluminescence [41].   It is well known that surface chalking and erosion
is a major problem in the application of epoxy resins as coatings
outdoors, and this technique provides a nondestructive test for the
incipient breakdown of the network that precedes microcrack formation.

E.   Moisture Effects

Many polymers absorb substantial amounts of water during ambient
environmental exposure and the properties of the polymer often
depend on moisture content.   In addition to changes in T_g, tensile
modulus, and strength, which may be reversible [58,59], there can
be permanent changes to polymer structure such as hydrolytic chain
scission and stress crazing produced by increases in internal stress
resulting from the changes in free volume that accompany the sorp-
tion and desorption of water [58,60,61].   This is of particular importance in matrices such as epoxy resins for coatings or aerospace
composites [62].   If the moisture produces a change in polymer free
volume, rate of oxidation, or free-radical production, then this may
be detectable by changes in the chemiluminescence properties.
      One obvious limitation in the use of chemiluminescence in studying
moisture effects on free volume is that the design of the experiment

should ensure that the method of measuring chemiluminescence does not alter the water content of the resin. Consequently, steady-state or nonstationary experiments that involve high temperatures cannot be used. The UV irradiation technique is ideally suited for these studies. Figure 14 shows the change in radical half-life τ, measured from the decay of chemiluminescence at 25°C from an aerospace composite resin (MY720-DDS as described in Section IV.D) as a function of the time that the block resin was immersed in water at 25°C [53]. There is an immediate decrease in τ with immersion that continues for up to 300 h and apparently levels off. However, there is a slow further decrease in τ after 40 days as can be seen from the point in Figure 14. The plasticization of the resin that causes the reported decrease in T_g [58] results in an increase in the radical mobility and thus a shortening of the half-life as measured by chemiluminescence. The rate of this plasticization as measured by the radical lifetime depends on the original crosslink density, which may

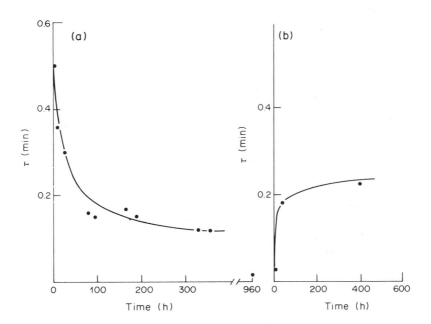

FIGURE 14   The effect of moisture on the radical lifetime τ from nonstationary chemiluminescence experiments on an epoxy resin (MY720 cured with 27% DDS). (a) Sample immersed in water at 25°C for the time indicated before measuring the UV stimulated chemiluminescence in air. (b) Sample immersed 40 days in water and then heated in vacuum at 60°C before measurement of UV chemiluminescence in air. (From Ref. 53.)

be controlled by the hardener concentration. If the resin, after 40 days immersion, is heated at 60°C under vacuum for different times prior to measurement of radical lifetime, there is seen to be an increase in τ as the water is removed (Figure 14). This agrees with the reversible plasticization of some epoxy resins [59] but it is seen that the lifetime does not return to its original value, suggesting that some permanent change to the network structure may have occurred.

Stress-induced chemiluminescence has been used as an alternative perturbation method for studying moisture effects [63]. On application of a stress of 45 MPa to a dry MY720-DDS epoxy resin, a steadily increasing chemiluminescence signal was obtained. When the stress was removed, the signal decayed over 20 min to a value only about one-half of that when stress was applied. When the experiment was repeated on a sample containing 0.6 wt% water, the signal immediately began to decay from a higher intensity once stress was applied, and the removal of stress resulted in a burst of emission followed by decay of the chemiluminescence to a zero emission intensity. Unfortunately, no mathematical analysis of the decay data was presented, but the results were attributed to the plasticizing action of the water. These observations are consistent with the results from the UV-stimulated chemiluminescence from the same resin system described above.

In a study of the chemiluminescence from cellulose fibers, permanent effects to the polymer of water exposure have been observed [64]. When a carefully controlled paper sample was subjected to humidity cycling by passing wet and dry air alternately over the sample, there was an increase in chemiluminescence intensity after each cycle. This was believed due to internal stresses on the fibers of the paper, resulting in free-radical production by chain scission of the cellulose. Below 40°C, the emission intensity in moist air was greater than in dry air. This is in agreement with nonstationary experiments in these laboratories on nylon 6 as a function of moisture content in which chemiluminescence intensity increased by a factor of 2. This could reflect the increase in k_6 due to increased free volume. The reported results of the chemiluminescence from paper were related to the well-known process of aging and embrittlement of paper under long-term storage. Scission of polymer chains in the amorphous region allows recrystallization of the smaller fragments, leading to embrittlement.

V. CONCLUSION

From an examination of the reported steady-state and nonstationary chemiluminescence experiments on commercial polymers in both inert and oxidizing atmospheres, it is concluded that the luminescence data may be most readily interpreted as resulting from the bimolecular termination of alkyl peroxy radicals which, in most cases, are produced from the decomposition of trace hydroperoxides. Such a chemiluminescence mechanism allows a kinetic scheme to be established that will describe the decay of emission produced after perturbation of the steady state by UV irradiation, gas switching, or mechanical stress. This technique offers more promise for the generation of kinetic data under nearly ambient conditions than do steady-state experiments in which the intensities become vanishingly small and induction periods prohibitively long. The following applications of chemiluminescence under nearly ambient conditions have been considered:

1. The study of the mechanisms of thermooxidative degradation of polyamides and polyolefins, the role of long-lived alkyl radicals, and the mode of action of stabilizers
2. The action of a hindered amine stabilizer and its nitroxyl radical during the UV degradation of polypropylene
3. The use of steady-state chemiluminescence to measure the T_g of epoxy resins and phase separation of blends
4. The monitoring of the viscosity of epoxy resins through the gel point by both steady-state and nonstationary experiments
5. The study of the chain scission and resulting change in free volume of crosslinked networks on UV irradiation
6. The effect of moisture on the structure of crosslinked networks and fibers

It is concluded that while chemiluminescence is extremely sensitive, to achieve its full potential it must first be coupled with other spectroscopic and analytical techniques to determine the details of the chemiluminescence process. It may then be extended to lower temperatures with the ultimate goal of being able to predict the performance of the polymer in service.

ACKNOWLEDGMENTS

The financial assistance of the Australian Research Grants Scheme and the Defence Science and Technology Organization for studies of polymer chemiluminescence is gratefully acknowledged.

REFERENCES

1. G. D. Mendenhall, Angew Chem. Int. Ed., 16: 225 (1977).
2. R. E. Kellogg, J. Am. Chem. Soc., 91: 5433 (1969).
3. G. E. Ashby, J. Polym. Sci., 50: 99 (1961).
4. G. A. George, Developments in Polymer Degradation, Vol. 3 (N. Grassie, ed.), Applied Science, London, p. 173 (1981).

5. E. S. O'Keefe and N. S. Billingham, Polym. Deg. Stabil., 10: 137 (1985).

6. N. M. Emanuel, G. E. Zalkov, and Z. K. Maizus, Oxidation of Organic Compounds: Medium Effects in Radical Reactions, Pergamon, London, p. 74 (1984).

7. G. A. George, G. T. Egglestone, and S. Z. Riddell, Polym. Eng. Sci., 23: 412 (1983).

8. M. P. Schard, Polym. Eng. Sci., 5: 246 (1965).

9. M. P. Schard and C. A. Russell, J. Appl. Polym. Sci., 8, 985:997 (1964).

10. L. Matisova-Rychla, J. Rychly, and M. Vavrekova, Eur. Polym. J., 14: 1033 (1978).

11. G. A. George and D. P. Schweinsberg, J. Appl. Polym. Sci., 33: 2281 (1987).

12. D. L. Fanter, C. J. Wolf, and K. M. Thiele, Polym. Mater. Sci. Eng., 54: 695 (1986).

13. V. A. Belyakov and R. F. Vassil'ev, Photochem. Photobiol., 11: 179 (1970).

14. G. A. George, Polym. Deg. Stabil., 1: 217 (1979).

15. G. D. Mendenhall, R. A. Nathan, and M. A. Golub, Applications of Polymer Spectroscopy (E. G. Brame, ed.), Academic Press, New York, p. 101 (1978).

16. R. F. Vassil'ev, Makromol. Chem., 126: 231 (1969).

17. L. Zlatkevich, Polym. Eng. Sci., 24: 1421 (1984).

18. M. Nikano, K. Takayama, Y. Shimizu, Y. Tsuji, H. Inaba, and T. Migita, J. Am. Chem. Soc., 98: 1974 (1976).

19. D. R. Kearns, Chem. Rev., 71: 395 (1971).

20. J. Turkevich, D. Mickewich, and G. T. Reynolds, Z. Phys. Chem., 82: 185 (1972).

21. R. A. Lloyd, Trans. Faraday Soc., 61: 2182 (1965).

22. L. Reich and S. Stivala, Makromol. Chem., 103: 74 (1967).

23. A. Garton, D. J. Carlsson, and D. M. Wiles, Makromol. Chem., 181: 1841 (1980).

24. P. S. Nangia and S. W. Benson, Int. J. Chem. Kinet., 12: 29 (1980).

25. F. R. Mayo, Macromolecules, 11: 942 (1978).

26. J. R. MacCallum, Developments in Polymer Degradation, Vol. 4 (N. Grassie, ed.), Applied Science, London, p. 191 (1984).

27. L. Matisova-Rychla, Z. Fodor, and J. Rychly, Polym. Deg. Stabil., 3: 371 (1980-81).

28. N. V. Zolotova and E. T. Denisov, J. Polym. Sci. A-1, 9: 3311 (1971).

29. N. C. Billingham and E. S. O'Keefe, "Oxyluminescence Studies of Polyamides and Polydienes," Proc. IUPAC Polymer 85, Melbourne, Australia, pp. 533—535 (1985).

30. E. T. Denisov, Developments in Polymer Stabilization, Vol. 5 (G. Scott, ed.), Applied Science, London, p. 23 (1982).

31. V. Ya Shlyapintokh, Russ. Chem. Revs., 35: 292 (1966).

32. V. Ya Shlyapintokh, O. N. Karpukhin, L. M. Postnikov, V. F. Tsepalov, A. A. Vichutinskii, and I. V. Zakharov, Chemiluminescence Techniques in Chemical Reactions, Consultants Bureau, New York, p. 164 (1968).

33. N. C. Billingham and E. S. O'Keefe, Chemiluminescence from Oxidative Degradation of Polymers, Project Report, University of Sussex, 1985.

34. D. J. Carlsson, A. Garton, and D. M. Wiles, Macromolecules, 9: 695 (1976).

35. B. A. Lloyd, K. L. DeVries, and M. L. Williams, J. Polym. Sci. A-2, 10: 1415 (1972).

36. G. A. George, G. T. Egglestone, and S. Z. Riddell, J. Appl. Polym. Sci., 27, 3999 (1982).

37. L. Zlatkevich, J. Polym. Sci., Polym. Phys. Ed., 23: 1691 (1985).

38. P. Vink, J. Appl. Polym. Sci., Appl. Polym. Symp., 35: 265 (1976).

39. F. R. Mayo, Polym. Preprints, 18: 402 (1977).

40. F. Yoshii, T. Sasaki, K. Makuuchi, and N. Tamura, J. Appl. Poly,. Sci., 30: 3339 (1985).

41. F. Yoshii, T. Sasaki, K. Makuuchi, and N. Tamura, J. Appl. Polym. Sci., 31: 1343 (1986).

42. W. K. Fisher, J. Indust. Irradiat. Tech., 3: 167 (1985).

43. R. J. Flemings and J. Hagekyriakau, Radiat. Prot. Dosim., 8: 99 (1984).

44. Z. Osawa, F. Konoma, S. Wu, and J. Cen, Polym. Photochem., 7: 33 (1986).

45. L. Matisova-Rychla, P. Ambrovic, N. Kulickova, J. Rychly, and J. Holcik, J. Polym. Sci., 57: 181 (1976).

46. T. V. Pokholok, O. N. Karpukhin, and V. Ya Shlyapintokh, J. Polym. Sci., Polym. Chem. Ed., 13: 525 (1975).

47. D. J. Carlsson, K. H. Chan, J. Durmis, and D. M. Wiles, J. Polym. Sci., Polym. Chem. Ed., 20: 575 (1982).

48. G. D. Mendenhall, J. A. Hassell, and R. A. Nathan, J. Polym. Sci., Polym. Chem. Ed., 15: 99 (1977).

49. R. A. Nathan, G. D. Mendenhall, M. A. Birts, C. A. Ogle, and M. A. Golub, ACS Adv. Chem. Ser., 169: 19 (1978).

50. V. Ya Shlyapintokh and V. B. Ivanov, Developments in Polymer Stabilization, Vol. 5, (G. Scott, ed.), Applied Science, London, p. 41 (1982).

51. D. K. C. Hodgeman, Developments in Polymer Degradation, Vol. 4 (N. Grassie, ed.), Applied Science, London, p. 189 (1982).

52. G. A. George, Pure Appl. Chem., 57: 945 (1985).

53. G. A. George and D. P. Schweinsberg, Corros. Sci., 26: 331 (1986).

54. K. Naito and T. K. Kwei, J. Polym. Sci., Polym. Chem. Ed., 17: 2935 (1979).

55. G. A. George, Mater. Forum, 9: 224 (1986).

56. J. Mijovic and J. A. Koutsky, Polymer, 20: 1095 (1979).

57. V. Bellenger and J. Verdu, J. Appl. Polym. Sci., 28: 2677 (1983).

58. R. J. Morgan, J. E. O'Neal, and D. L. Fanter, J. Mater. Sci., 15: 751 (1980).

59. M. K. Antoon, J. L. Koenig, and T. Serafine, J. Polym. Sci., Polym. Phys. Ed., 19: 1567 (1981).

60. M. K. Antoon and J. L. Koenig, J. Polym. Sci., Polym. Phys. Ed., 19: 197 (1981).

61. R. L. Levy, D. L. Fanter, and C. J. Summers, J. Appl. Polym. Sci., 24: 1643 (1979).

62. J. Mijovic and K. Lin, J. Appl. Polym. Sci., 30: 2527 (1985).

63. R. L. Levy and D. L. Fanter, Polym. Preprints, 20: 543 (1979).

64. G. B. Kelly, J. C. Williams, G. D. Mendenhall, and C. A. Ogle, Durability of Macro Molecular Materials (R. K. Eby, ed.), American Chemical Society, Washington, D.C., p. 117 (1979).

65. G. D. Mendenhall, M. Ngo, and H. K. Agarwal, J. Appl. Polym. Sci., 33: 1259 (1987).

66. L. Zlatkevich, Radiothermoluminescence and Transitions in Polymers, Springer-Verlag, New York, p. 23 (1987).

4
Chemiluminescence in Evaluating Thermal Oxidative Stability.

LEV ZLATKEVICH   VIBO Research, Inc., Pennsauken, New Jersey

Cab. p 192

I. INTRODUCTION

Each year brings us an increasing demand for polymeric materials
and growing requirements regarding their physicomechanical characteristics.   Although many of the early applications were in areas where
materials were exposed to low mechanical and environmental stress,
recent applications have often been much more demanding.   The
question regarding material selection with the emphasis on its long-
term performance arises whenever a product is introduced or re-
designed.   Long-term performance can mean many things.   One of
the most important of those is the ability to resist the deteriorating
effects of the atmosphere, to which oxidation is a prominent contri-
butor.
      It has long been a goal of polymer users to predict the utility of
polymeric materials by short-term accelerated tests capable of describing long-term in-service performance.   Conventional laboratory weathering chambers usually employ harsher than natural exposure
conditions in order to shorten the required exposure time.   Fairly
long exposure times (500 to 2000 hr) may still be necessary.   Since
degradation and stabilization mechanisms which occur in an accelerated
test may not take place during use, persistent problems with utilizing
more severe testing conditions than those experienced during use
are to be expected.   Nonetheless, in recent years correlations have
been attempted, with reasonable success, between accelerated ultraviolet exposure procedures such as Xenon Weatherometer or Xenotest
and outdoor weathering in various climates [1].   Unfortunately, such

correlations are not readily obtainable for thermal oxidative aging because of the tremendously long time frames involved in some applications. For example, polyolefins used in the telecommunications industry for wire and cable jackets must be stable for periods of up to 40 years [2].

Among various methods which have been proposed and are being used to evaluate the oxidative stability, there is no ideal simulated test method. At best there is only a fair correlation between the methods. The more the test differs from actual environmental conditions, the less likely is the performance of the polymer expected to correlate with commercial usage and processing. The time-honored way of attempting to predict long-term performance of polymers is through use of an Arrhenius-type plot of the temperature dependence of some oxidation-sensitive property of the polymer system in terms of reciprocal absolute temperature against the logarithm of time. This approach has the obvious advantage of reducing the exponential relationship which seems best to describe the phenomenon to a straight-line approximation more amenable to direct graphical extrapolation. Frequently, however, such predictions are inaccurate since controlling new factors can enter at temperatures either above or below the testing temperature, factors which can completely vitiate extrapolation in either direction. The results obtained for some polyethylene samples are very illustrative: extrapolation of higher temperature data to 70°C produced an expected lifetime of about 200 years whereas actual aging at 70°C showed stability of only 100 days [3]. Quite obviously a new variable, not operative at test temperatures in the melt, appeared and assumed the controlling role.

The legitimate question may then be raised, is it possible at all to test long-term use sensibly in a laboratory? It is certainly possible to simulate different conditions in a testing apparatus. If, however, the aging conditions are really simulated according to the true sense of the word, then the test loses its accelerating power and may require an enormous time. This is not practical, since one can not wait for years in order to find out whether a new material is usable. Probably the most promising attempt to solve this dilemma would be to use the exposure conditions approaching as closely as possible the real aging conditions and at the same time be able to complete the test in a reasonably short time. However, the methods traditionally used for studying the kinetics of liquid phase oxidation are too insensitive to provide useful results with polymers at low temperatures. Thus there is a clear and established need for more sensitive techniques for probing oxidation kinetics and one possibility is the measurement of the light emission associated with degradation.

Chemiluminescence has attracted much attention from chemists because of its association with oxidation and its extreme sensitivity, which permits the study of oxidative reactions at very early stages—much earlier than those at which IR and calorimeteric techniques

become sufficiently effective. The emission of very low levels of
light which accompanies the oxidation of all organic materials has been
known and well documented for many years [4—8].

The first application of chemiluminescence measurements to polymers was reported in 1961 by Ashby [9], who showed that luminescence occurs from many polymers when they are heated in air and is
associated with oxidation. He also showed that the luminescence
intensity is suppressed by antioxidants and suggested that the measurements of the stationary levels of luminescence may be useful for
judging the thermooxidative stability of polymers. Similar studies
were reported by Schard and Russell [10,11], who also demonstrated
the connection between luminescence and oxidation. Further essential
developments in this field are connected with numerous studies under-
taken by George et al. [12—16], Mendenhall et al. [17—20], and
Matisova-Rychla et al. [21—24]. The reaction rate constants and the
temperature coefficients or apparent activation energies for light-
producing reactions occurring during different stages of oxidation
have been calculated for several polymeric systems and were found
to agree well with the available data obtained by other techniques.

There are two different types of chemiluminescence experiment
that are generally performed corresponding to stationary and nonstationary kinetic conditions [12]. The first is an experiment in which
the light emitted by the sample is measured as a function of time of
thermal oxidation. The second approach is to perturb the environment of the polymer and measure the change in chemiluminescence
intensity as the material responds to the perturbation and then
returns to the steady state. The perturbation methods which are
usually explored include UV irradiation, mechanical stress, or changing the atmosphere above the polymer from inert to oxidizing. Nonstationary and stationary experiments differ significantly in the time
required to complete the test. The duration of the former is relatively short for both unstabilized and stabilized polymers, whereas the
latter may take quite a long time especially when applied to a highly
stabilized system. In spite of the time-saving nature of the nonstationary experiments, their applicability for evaluation of polymer
long-term thermal oxidative stability on a routine basis remains to be
shown. On the other hand, the stationary experiments provide
information on long-term polymer oxidation, and thus they will be
the subject of our discussion.

II. OXIDATION MECHANISM

It has been known for many years that the number of reactions
occurring during the thermal oxidation of polymers is very large.
By the late 1960s, however, it was generally accepted that the critical primary and secondary reactions observed in the oxidation of

liquid hydrocarbons applied to the oxidation of solid polyolefins as
well. Moreover, the same basic reactions seemed to be applicable to
polymers in general. Although there is still no detailed understanding of the oxidation of solid polymers, there is a considerable merit
in refocusing attention on the simplified, universal scheme for hydrocarbon oxidation described in terms of the initiation, propagation,
and termination reactions:

$$\text{Initiator} \xrightarrow{k_1} 2R^{\cdot} \qquad \text{Initiation} \qquad (1)$$

$$R^{\cdot} + O_2 \xrightarrow{k_2} RO_2^{\cdot} \qquad\qquad\qquad (2)$$

$$\qquad\qquad\qquad\qquad\qquad \text{Propagation}$$

$$RO_2^{\cdot} + RH \xrightarrow{k_3} ROOH + R^{\cdot} \qquad (3)$$

$$R^{\cdot} + R^{\cdot} \xrightarrow{k_4} R{-}R \qquad\qquad\qquad (4)$$

$$R^{\cdot} + RO_2^{\cdot} \xrightarrow{k_5} ROOR \qquad \text{Termination} \qquad (5)$$

$$RO_2^{\cdot} + RO_2^{\cdot} \xrightarrow{k_6} ROOR + O_2 \qquad (6)$$

The question of initiation was the subject of some controversy
and, indeed, remains a matter of opinion to this day. Nonetheless,
there is considerable evidence that all polymers which have been
thermally processed contain some macrohydroperoxides. In view of
the thermal instability of these groups, it seems likely that their
decomposition will be a major factor in initiating the oxidative degradation of polymers subjected to high temperatures. Bolland and Gee
[25] were probably the first to recognize the significance of macrohydroperoxides in oxidation of polymers. Energetically bimolecular
decomposition of hydroperoxides is favorable and the initiation reaction is considered to be

$$2ROOH \xrightarrow{k_1} RO_2^{\cdot} + RO^{\cdot} + H_2O \qquad (7)$$

At very low hydroperoxide concentrations, when molecular association
of two hydroperoxide molecules has low probability, the unimolecular
hydroperoxide decomposition is also considered:

$$ROOH \xrightarrow{k_1'} RO^{\cdot} + {}^{\cdot}OH \qquad (8)$$

III. CHEMILUMINESCENCE MECHANISM

A. Chemiluminescence Under Oxygen Atmosphere

One problem in the interpretation of luminescence data has been that there is no general agreement on the origin of luminescence. It is generally accepted that the emission comes from phosphorescence of an excited ketone but opinions differ on the origin of the ketone. Most workers believe that it is produced in the termination reaction of two alkyl peroxy radicals via the mechanism:

$$R-CH_2-OO^{\cdot} + {}^{\cdot}OOR \longrightarrow R-\overset{\displaystyle *}{\underset{\displaystyle H}{C}}=O + ROH + O_2 \qquad (9)$$

If this is true, then the intensity of light is expected to be proportional to the rate of termination, so that

$$I = C[RO_2^{\cdot}]^2 \qquad (10)$$

This model seems to work well for liquid phase oxidation and has been used widely.

In contrast, other workers have suggested that the luminescence is produced by the decomposition of the polymer hydroperoxides. In this case luminescence arises either from a molecular rearrangement reaction:

$$R-CH_2-OOH \longrightarrow R-\overset{\displaystyle *}{\underset{\displaystyle H}{C}}=O + H_2O \qquad (11)$$

or from a two-stage process in which disproportionation of the products of homolysis of the O–O bond is favored by the cage entrapment of the radicals in the polymer matrix:

$$R-CH_2-OOH \longrightarrow RCH_2O^{\cdot\cdot}OH \longrightarrow R-\overset{\displaystyle *}{\underset{\displaystyle H}{C}}=O + H_2O \qquad (12)$$

In either case we would expect to find that the luminescence intensity is proportional to the hydroperoxide concentration:

$$I = C[ROOH] \qquad (13)$$

Both mechanisms (9) and (11, 12) have been utilized but neither of them can be claimed to be fully satisfactory. The major problem with the mechanism (9) is that it requires at least one of the combining radicals to be primary or secondary, the condition which is not satisfied in some cases (e.g., with polypropylene). Since the main sites of the oxidation of polypropylene are the tertiary C—H bonds, one would expect no chemiluminescence from this polymer. In fact, it is well known that polypropylene does luminesce and indeed is more emissive than polyethylene.

The mechanism (11, 12) is not flawless either, since the acceptance of either of reactions (11) or (12) means that hydroperoxide decomposition produces light but fails to produce radicals for chain initiation.

A third mechanism recently proposed by Audouin-Jirackova and Verdu [27] links the chemiluminescence to the β scission of alkoxy radicals via the formation of a transient biradical:

$$
\begin{array}{c}
CH_3 \\
| \\
-C-CH_2- \\
| \\
O^\cdot
\end{array}
\longrightarrow
\left[
\begin{array}{c}
^\cdot CH_2- \\
+ \\
CH_3 \\
| \\
-C^\cdot \\
| \\
O^\cdot
\end{array}
\right]
\longrightarrow
\left[
\begin{array}{c}
O \\
\| \\
-C-CH_3
\end{array}
\right]^*
\longrightarrow
\begin{array}{c}
O \\
\| \\
-C-CH_3 + h\nu
\end{array}
\tag{14}
$$

The latter mechanism has to be distinguished from one discussed by Mendenhall et al. [41] whereby the luminescent ketone results from the disproportionation of two alkoxy radicals and whereby the reaction is blocked by tertiary substitution. It also differs from the mechanism introduced in the previous chapter whereby the β scission of alkoxy radicals is utilized as the intermediate step in transformation of tertiary radicals to primary radicals.

Audouin-Jirackova and Verdu's mechanism seems to be superior to all other mechanisms because it is applicable to substances with tertiary carbon atoms and also because the decomposition of hydroperoxides in this case accounts for both initiation of oxidation and the emission of light:

$$
ROOH \xrightarrow{k_j} HO^\cdot + RO^\cdot
$$

$$
\text{(14) route} \longrightarrow h\nu
$$

$$
\text{chain transfer with RH} \longrightarrow R^\cdot
\tag{15}
$$

Although we intend to give the preference to the chemiluminescence mechanism based on hydroperoxide decomposition, it is worthwhile to mention that both mechanisms represented by Equation (10) and Equation (13) are equivalent when applied to stationary chemiluminescence experiments; the fact first noted by Vasil'ev [62]. This is the consequence of the equality of the rates of initiation and termination for a linear steady state chain reaction. Thus the long standing question regarding the chemiluminescence mechanism is irrelevant to the stationary experiments and in the following sections the kinetic relationships for chemiluminescence generated at stationary conditions will be developed by utilizing the relation $I = C[ROOH]$.

B.   Chemiluminescence Under Inert Atmosphere

The peroxy radical recombination as the cause of chemiluminescence may not be applied in the absence of oxygen. Thus it is certain that hydroperoxide decomposition is involved. Along with the possibilities stated above [reactions (11), (12), and (15)], the hydroperoxide decomposition in a bimolecular reaction with the participation of chemisorbed oxygen was discussed by Matisova-Rychla et al. [22]:

$$R-CH_2-\underset{\underset{OOH \cdot O_2^-}{|}}{\overset{\overset{CH_3}{|}}{C}}-CH_2-\overset{\overset{CH_3}{|}}{CH}- \xrightarrow{(e^-)} H_2O_2 + R-CH_2-OH + CH_3-\underset{\underset{O}{\|}}{\overset{\overset{CH_3}{|}}{C}}-CH=C-$$

$$(16)$$

In the above scheme an O_2^- radical ion is formed in the presence of suitable electron donors as the result of oxygen electron affinity.

IV.   OXIDATION KINETICS AND CHEMILUMINESCENCE

A.   Chemiluminescence Under Oxygen Atmosphere

1.   Initiation by Bimolecular Hydroperoxide Decomposition: Initial Stages of Oxidation

Under mild conditions of oxidation the chain lengths are long and the amount of oxygen participating in the reaction approximately equals the amount of hydroperoxides formed. Under this condition the amount of hydroperoxide which decomposes to initiate further oxidation is very small and its absence from the expression for oxidation rate is valid. At high oxygen pressures, steps (4) and (5) can be neglected and the solution of Equations (1) through (6) using the steady-state approximation is

$$-\frac{d[O_2]}{dt} = \frac{d[ROOH]}{dt} = k_3\left(\frac{k_1}{k_6}\right)^{1/2}[ROOH][RH] \qquad (17)$$

For long chain lengths, many molecules of hydroperoxide are formed per free radical initiating the reaction before termination occurs and hence variations of overall rate constant $k = k_3(k_1/k_6)^{1/2}$ essentially reflects changes in the rate of propagation.* If one denotes polymer and hydroperoxide concentrations by [A] and [B], respectively, Equation (17) can be rewritten as

$$\frac{d[B]}{dt} = k[A][B]$$

(18)

Equation (18) presents the autocatalytic reaction with regard to substrate and hydroperoxide and, as it is typical for autocatalytic reactions, induction and acceleration periods should be expected. Designating the increase in [B[during the oxidation as $X = [B] - [B]_0$ and noting that the increase in [B] is equal to the decrease in [A], $[B] - [B]_0 = [A]_0 - [A]$,

$$\frac{dX}{dt} = k([A]_0 - X)([B]_0 + X)$$

(19)

where $[A]_0$ and $[B]_0$ are the initial polymer and hydroperoxide concentrations, respectively. Integration of Equation (19) gives

$$k([A]_0 + [B]_0)t = \ln\left(\frac{[B]_0 + X}{[A]_0 - X} \cdot \frac{[A]_0}{[B]_0}\right)$$

(20)

Since the chemiluminescence emission intensity is proportional to hydroperoxide concentration [see Equation (13)]:

$$I_t = C[B] = C([B]_0 + X)$$

(21)

When the chemiluminescence intensity reaches the maximum,

$$I_{max} = C([A]_0 + [B]_0) \quad \text{or} \quad I_{max} = C([A]_0 - X + [B]_0 + X)$$

(22)

Substituting $([B]_0 + X)$ and $([A]_0 - X)$ from Equations (21) and (22) into Equation (20) and taking into account that in all practical cases $[A]_0 \gg [B]_0$,

$$\ln\left(\frac{I_t}{I_{max} - I_t}\right) = \ln\frac{[B]_0}{[A]_0} + k[A]_0 t$$

(23)

---

*I wish to express my appreciation to Drs. Billingham and George, who brought to my attention the fact that the replacement of the overall reaction rate constant k by the propagation rate constant k_3 is not appropriate.

Equation (23) provides a convenient method of estimating $\ln([A]_0/[B]_0)$ and $k[A]_0$ values. The first expression is proportional to induction time, whereas the second is proportional to oxidation rate. A plot of $\ln[I_t/(I_{max} - I_t)]$ against t has slope $k[A]_0$ and intercept $\ln([B]_0/[A]_0)$.

Although the direct relationship between oxidation rate and $k[A]_0$ is straightforward and does not require further elucidation[*], the proportionality between the induction time and $\ln([A]_0/[B]_0)$ is less obvious and may need substantiation. Oxidation is a radical chain reaction. The theory of chain reactions predicts that in the initial stages of any self-accelerating reaction the amount X of the substance having undergone reaction varies with the time t according to the law [26]:

$$X = C(e^{\psi t} - 1) \qquad (24)$$

where C and ψ are constants. The amount of X can be detected with the aid of the apparatus of a given sensitivity only after a time interval τ (called the "induction time") when it has already attained a definite value X_0. In terms of our nomenclature, Equation (24) can be replaced by its equivalent:

$$[B]/[B]_0 = e^{\psi t} \qquad (25)$$

Thus the induction time decreases exponentially with the increase in the initial hydroperoxide concentration and can be expressed as $\ln([A]_0/[B]_0)$. It has to be underlined that the advantage of the evaluation according to Equation (23) is that the results do not depend on the amount of the material analyzed, and the induction time and oxidation rate values obtained for various samples of different weight can be directly compared. However, if the weight of the sample, or more correctly, the surface area of the samples emitting light is kept constant, Equation (23) can be simplified and only the initial exponential portion of the curve need be utilized. When $I_t \ll I_{max}$,

$$\ln I_t = \ln(C[B]_0) + k[A]_0 t \qquad (26)$$

In Equation (26) $-\ln(C[B]_0)$ is proportional to induction time, whereas oxidation rate is expressed similarly to Equation (23) by $k[A]_0$.

It is worthwhile to note that the equation of the same type as Equation (23) can be deduced if initially there are no hydroperoxides

---

[*]Taking into account that $[A]_0 \gg [B]_0$, Equation (18) can be presented as $d[B]/dt \simeq k[A]_0[B] = k'[B]$. The latter equation refers to a pseudo-first order reaction with the reaction rate constant $k' = k[A]_0$.

in the system ($[B]_0 = 0$) but the oxidation reaction consists of the two parallel steps: a noncatalytic first order transformation and an autocatalytic transformation of A into B. In this case Equation (18) is replaced by its equivalent:

$$\frac{d[B]}{dt} = k[A][B] + k_0[A] \qquad (27)$$

or

$$\frac{dX}{dt} = k([A]_0 - X)(X + k_0/k) \qquad (28)$$

where k_0 is the rate of initiation.

Equation (28) is formally the same as Equation (19) where $[B]_0$ is replaced by k_0/k. Thus all conclusions reached previously for oxidation initiated by bimolecular hydroperoxide decomposition are also applicable to the case when hydroperoxides are originally not present in the system but appear as a result of some external influence such as heat, photolysis, high-energy irradiation, etc. In this case Equation (23) can be presented in a slightly different form:

$$\ln\left(\frac{I_t}{I_{max} - I_t}\right) = \ln\frac{k_0}{k[A]_0} + k[A]_0 t \qquad (29)$$

where the parameters to be evaluated are the oxidation rate constant $k[A]_0$ and the initiation rate constant k_0.

2. Initiation by Bimolecular Hydroperoxide Decomposition: Advanced Stages of Oxidation

Under relatively severe conditions of oxidation (high temperatures, long time intervals, presence of metallic activators and light) the decomposition of the hydroperoxides becomes appreciable, and the rate of oxidation can no longer be equated with the rate of hydroperoxide formation as represented by the first two terms in Equation (17). In this case the disappearance of hydroperoxides by Equation (7) should be included in the equation for the rate of change of hydroperoxide concentration with time:

$$\frac{d[ROOH]}{dt} = k_3\left(\frac{k_1}{k_6}\right)^{1/2}[ROOH][RH] - k_1[ROOH]^2 \qquad (30)$$

The advanced stages of oxidation must be marked by appreciable disappearance of substrate and this factor may be introduced into

the above equation, i.e., $[RH] = [RH]_0 - ([ROOH] - [ROOH]_0)$.
This leads to the following equation:

$$\frac{d[ROOH]}{dt} = k_3 \left(\frac{k_1}{k_6}\right)^{1/2} [ROOH]([RH]_0 + [ROOH]_0)$$

$$- \left[k_3 \left(\frac{k_1}{k_6}\right)^{1/2} + k_1\right] [ROOH]^2 \qquad (31)$$

Equation (31) can be integrated to give

$$[ROOH] = \frac{[ROOH]_\infty}{1 - \left(1 - \dfrac{[ROOH]_\infty}{[ROOH]_0}\right) e^{-at}} \qquad (32)$$

where $[ROOH]_\infty$ is the steady-state value of hydroperoxide (the concentration approached at long times as $t \longrightarrow \infty$) and $[ROOH]_0$ is the initial concentration of hydroperoxides; $a = k_3(k_1/k_6)^{1/2}([RH]_0 + [ROOH]_0) = k([RH]_0 + [ROOH]_0)$, and $[ROOH]_\infty = a/[k_3(k_1/k_6)^{1/2} + k_1] = (k/k + k_1)([RH]_0 + [ROOH]_0)$. As it follows from Equation (32), the concentration of hydroperoxides approaches a steady-state value which is a maximum value if $[ROOH]_0 < [ROOH]_\infty$ and a minimum value if $[ROOH]_0 > [ROOH]_\infty$.
Three situations can exist:

(1) $[ROOH]_0 \ll [ROOH]_\infty$
Replacing $[RH]_0$, $[ROOH]_0$, $[ROOH]$, and $[ROOH]_\infty$, by $[A]_0$, $[B]_0$, $[B]$ and $(k/k + k_1)([A]_0 + [B]_0)$

$$[B] = \frac{(k/k + k_1)([A]_0 + [B]_0)}{1 - \left[1 - \dfrac{(k/k + k_1)([A]_0 + [B]_0)}{[B]_0}\right] e^{-k([A]_0 + [B]_0)t}} \qquad (33)$$

Since $I_t = C[B]$ and $I_{max} = C(k/k + k_1)([A]_0 + [B]_0)$

$$\ln\left(\frac{I_t}{I_{max} - I_t}\right) = \ln\left[\frac{[B]_0}{(k/k + k_1)([A]_0 + [B]_0) - [B]_0}\right] + k([A]_0 + [B]_0)t \qquad (34)$$

Eq. (34) resembles Eq. (23) and it is certainly of interest to compare them in more detail. The latter was deduced for autooxidation reactions with long chain lengths, whereas the former was developed without any preconditions regarding the reaction chain length. Since $[A]_0 \gg [B]_0$

(this condition is fulfilled in all practical cases dealing with autooxidation),
both equations have the same second term on the right-hand side. The
oxidation chain length ν can be expressed as $\nu = (k/k_1)([A]/[B])$
Provided the conversion of autooxidation is not much different from
50%, one may consider two extreme possibilities:

1. ν is very large. Then $k \gg k_1$ and Eq. (34) transforms into
Eq. (23).

2. ν approaches 1. Since $[B]_0$ is typically at least 3-4 orders of
magnitude smaller than $[A]_0$ while k and k_1 values are similar (vary
at most by one order of magnitude), the difference between Eqs. (23)
and (34) reduces to the difference between $\ln([B]_0/[A]_0)$ and
$\ln([B]_0/[A]_0) + \ln[(k+k_1)/k)]$. The term $\ln[(k + k_1)/k]$ modifies
the total $\ln[k + k_1)/k]$ value by no more than 10 to 15 percent, which
is within the experimental accuracy of measurements. This justifies
the applicability of Eq. (23) to the oxidation chain reactions irrespective of the reaction chain length.

(2) $[ROOH]_0 < [ROOH]_\infty$

$I_t = c[B]$, $I_0 = C[B]_0$, $I_{max} = C(k/k + k_1)([A]_0 + [B]_0)$

$$\ln\left[\frac{I_t(I_{max} - I_0)}{I_0(I_{max} - I_t)}\right] = k([A]_0 + [B]_0)t \tag{35}$$

(3) $[ROOH]_0 > [ROOH]_\infty$

$I_t = C[B]$, $I_0 = C[B]_0$, $I_{min} = C(k/k + k_1)([A]_0 + [B]_0)$

$$\ln\left[\frac{I_t(I_0 - I_{min})}{I_0(I_t - I_{min})}\right] = k([A]_0 + [B]_0)t \tag{36}$$

3. Initiation by Unimolecular Hydroperoxide Decomposition: Initial
Stages of Oxidation

Utilizing the steady-state approximation, the solution for the Equations
(1) to (6) is

$$-\frac{d(O_2)}{dt} = \frac{d[ROOH]}{dt} = k[ROOH]^{1/2}[RH] \tag{37}$$

or using previous nomenclature

$$\frac{dX}{dt} = k([A]_0 - X)([B]_0 + X)^{1/2} \tag{38}$$

Integrating Equation (38) and utilizing the series $\ln[(1 + x)/(1 - x)] \simeq 2x$ at $|x| \ll 1$, one gets

$$k\sqrt{[A]_0 + [B]_0}\, t = 2\left(\sqrt{\frac{[B]_0 + X}{[A]_0 + [B]_0}} - \sqrt{\frac{[B]_0}{[A]_0 + [B]_0}}\right) \qquad (39)$$

Taking into account that $[A]_0 \gg [B]_0$, $I_t = C([B]_0 + X)$ and $I_{max} = C([A]_0 + [B]_0)$ and assuming that the contribution to luminescence from the originally present hydroperoxides is negligibly small

$$\frac{I}{t} = \frac{k^2 C\,[A]_0^2}{4}\, t \qquad (40)$$

Thus plotting of I/t as a function of t yields a straight line with the slope $k^2 C\,[A]_0^2/4$ from which the value proportional to the oxidation rate can be evaluated.

B. Chemiluminescence Under Inert Atmosphere

1. Unimolecular Hydroperoxide Decomposition

In the absence of oxygen the change in the chemiluminescence intensity with time or temperature is not directly related to oxidation. Nonetheless, important information concerning the hydroperoxides present in the system can be obtained.

For the first order unimolecular hydroperoxide decomposition,

$$\frac{d[ROOH]}{dt} = -k_D[ROOH] \qquad (41)$$

and upon integration,

$$[ROOH] = [ROOH]_0\, \exp(-k_D t) \qquad (42)$$

where k_D is the rate of decomposition.   Chemiluminescence intensity can be expressed as

$$I = Ck_D[ROOH] \qquad (43)$$

or

$$I = Ck_D[ROOH]_0\, \exp(-k_D t) = I_o\, \exp(-k_D t) \qquad (44)$$

Then the rate of isothermal unimolecular hydroperoxide decomposition k_D may be evaluated as the slope of the straight line plotted in $\ln(I_0/I)$ vs. t coordinates.

The decomposition of hydroperoxides can be studied under other than isothermal conditions.   When the sample is heated the hydroperoxide decomposition rate increases whereas the concentration of hydroperoxide taking part in the reaction decreases. Thus the number of photons emitted per second passes through a maximum due to these two competing effects (Figure 1). If temperature is increased at a constant rate $\beta = dT/dt$, it follows from Equations (41) and (43) that

$$I = -C\beta\frac{d[ROOH]}{dT} \tag{45}$$

Then

$$\int_{T_0}^{\infty} IdT = C\beta[ROOH]_0 \tag{46}$$

and

$$\int_{T}^{\infty} IdT = C\beta[ROOH] \tag{47}$$

Finally, from Equations (47), (45), and (41) we have

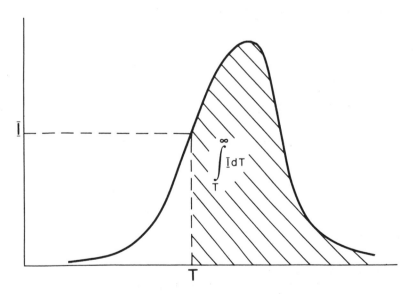

FIGURE 1   Chemiluminescence curve associated with the decomposition of hydroperoxides in an inert atmosphere.

$$k_D = \frac{I\beta}{\int_T^\infty IdT} \tag{48}$$

Thus for unimolecular hydroperoxide decomposition the total area under the curve is proportional to the initial hydroperoxide content (Equation (46)). Also, utilizing Equation (48), the rate of hydroperoxide decomposition at various temperatures can be evaluated.

2. Bimolecular Hydroperoxide Decomposition

In the case of second order bimolecular hydroperoxide decomposition we have

$$\frac{d[ROOH]}{dt} = -k_D'[ROOH]^2 \tag{49}$$

Integration of Equation (49) gives

$$\frac{1}{[ROOH]} - \frac{1}{[ROOH]_0} = k_D' t \tag{50}$$

Utilizing the relation $I/C = -d[ROOH]/dt = k_D'[ROOH]^2$ and converting to the chemiluminescence intensities,

$$\left(\frac{I_0}{I}\right)^{1/2} - 1 = k_D'[ROOH]_0 t \tag{51}$$

Although Equation (51) does not provide the ground for the calculation of the second order reaction rate constant k_D', it can be used for evaluation of the reaction half-life $(1/k_D'[ROOH]_0)$ if the results are plotted in $[(I_0/I)^{1/2} - 1]$ vs. t coordinates.

In the case of a nonisothermal experiment, the coefficient of proportionality (C) which relates the intensity of emitted light (I) and the rate of hydroperoxide decomposition $(d[ROOH]/dt)$ can be presented as

$$C = \frac{\int_{T_0}^\infty IdT}{\beta[ROOH]_0} \tag{52}$$

and the expression for the amount of hydroperoxides present at any instant is

$$[ROOH] = [ROOH]_0 \frac{\int_T^\infty I dT}{\int_{T_0}^\infty I dT} \tag{53}$$

Simple mathematical transformations lead to the equation

$$k_D'[ROOH]_0 = \frac{\beta I \int_{T_0}^\infty I dT}{\left(\int_T^\infty I dT\right)^2} \tag{54}$$

While Equations (44) and (51) provide the opportunity for the evaluation of the reaction order at isothermal conditions, Equations (48) and (54) can be used for the same purpose in constant heating rate experiments.

3. Evaluation of the Activation Energy of Hydroperoxide Decomposition

When the rate of hydroperoxide decomposition at different temperatures is known from either isothermal ot nonisothermal measurements, the activation energy of decomposition can be obtained. The activation energy plot (the plot of the reaction rate vs. 1/T) would yield a straight line if, and only if, the correct reaction order had been assumed.

The Arrhenius-type equations for the change in chemiluminescence intensity with temperature are

$$I = Ck_0[ROOH] \exp(-E/kT) \tag{55}$$

and

$$I = Ck_0' [ROOH]^2 \exp(-E/kT) \tag{56}$$

for the first and second order decompositions, respectively. At the beginning of the peak the hydroperoxide concentration changes only slightly with temperature, and therefore $I \propto \exp(-E/kT)$ for both first and second order reactions. This provides the basis for the evaluation of the reaction activation energy by the so-called method of initial rises which is independent of the reaction order. The plot of ln I as a function of 1/T yields a straight line the slope of which is $-E/k$.

For the first order kinetics, the temperature of the chemiluminescence maximum can be obtained by differentiation of Equation (55).

At the point of maximum the condition $d^2[ROOH]/dt^2 = 0$ should be fulfilled, which for a linear heating rate β gives (k = Boltzman constant)

$$\frac{k_0 kT^2_{max}}{E\beta} = \exp \frac{E}{kT_{max}} \tag{57}$$

As it follows from Equation (57), the position of the maximum does not depend on the hydroperoxide concentration, and thus the first order peak is not expected to shift on variation in the initial hydroperoxide concentration. Differently from the first order case, the increase in the initial hydroperoxide concentration should lead to the shift of the peak to lower temperatures if reaction follows second order kinetics [31].

In addition to the possibilities discussed previously, the activation energy of the first order hydroperoxide decomposition can be evaluated by utilization of Equation (57) since a plot of $\ln(T^2_{max}/\beta)$ against $1/T_{max}$ has slope E/k. Of course, such an evaluation would require the performance of the experiments at various heating rates.

C. Chemiluminescence Evaluation of the Long-Term Stability of Polymers

In the process of oxidation, the degree of conversion α changes with time t according to the relation:

$$\alpha = \frac{[X]}{[A]_0} = \frac{\dfrac{[B]_0}{[A]_0}\left(e^{k[A]_0 t} - 1\right)}{\dfrac{[B]_0}{[A]_0} e^{k[A]_0 t} + 1} \tag{58}$$

Equation (58) follows from Equation (20) when $[A]_0 \gg [B]_0$.

Let us define the critical degree of conversion α_c as the conversion above which material quality deteriorates to an unacceptable level. Then the lifetime of the material t_c can be expressed as

$$t_c = \frac{\ln\left(\dfrac{\alpha_c \cdot e^a - 1}{1 - \alpha_c}\right)}{b} \tag{59}$$

where $a = \ln([A]_0/[B]_0)$ and $b = k[A]_0$ are the induction time and oxidation rate, respectively. In terms the relative importance of the induction time and oxidation rate values it can be said that t_c is defined mostly by the induction time when $\alpha_c < 1/2$ and by the oxidation rate when $\alpha_c > 1/2$. In the particular case when $\alpha_c = 1/2$, $t_c \simeq a/b$.

The prediction of the material's lifetime t_c' at a temperature T_1 different from the temperature of the experiment T can be made by utilizing the activation energy of oxidation E. If $T_1 < T$,

$$t_c' = \frac{\ln\left(\dfrac{\alpha_c \cdot e^a - 1}{1 - \alpha_c}\right)}{b} \cdot e^{\frac{E}{k}(1/T_1 - 1/T)} \tag{60}$$

The knowledge of the α_c as well as careful evaluation of the activation energy of oxidation and the confidence in not crossing the intervals of major polymer temperature transitions are the necessary conditions for a meaningful extrapolation.

Equation (60) can be simplified if it is assumed that critical hydroperoxide concentration is reached when 50% of a substance is oxidised, i.e., $\alpha_c = 1/2$. Then

$$t_c' \simeq \frac{a}{b} \exp[E/k \ (1/T_1 - 1/T)] \tag{61}$$

The assumption $\alpha_c = 1/2$ is equivalent to the often used, although not justified, evaluation of the induction time as the time to reach one-half of maximum chemiluminescence intensity ($t_{1/2}$) (58).

V. INSTRUMENTATION

The basic instrumentation needed for chemiluminescence measurements consists of a heating arrangement for the sample in a light-tight box, a photomultiplier tube (PMT) to sense the chemiluminescence emission, an amplifier to measure the PMT output current, and a chart recorder to record this current as a function of temperature or time. Instrumentation used in chemiluminescence studies varies greatly in complexity. The simplest system can consist of a hot plate and chemical reaction flask in a light-tight box with a PMT and associated electronics. Complex apparatus could include a specially designed cell, sophisticated temperature and gas flow control, as well as an automated data aquisition and evaluation system.

For a long time there were no commercial instruments for the measurement of chemiluminescence. The lack of availability of equipment has certainly hampered the progress in the field and impel some researchers either to modify the existing equipment or construct their

a

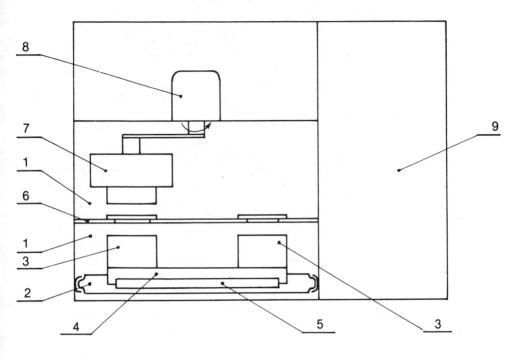

b

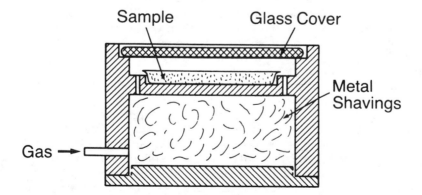

FIGURE 2   The scheme of the multisample chemiluminescence apparatus
(a) and the individual cell (b).   (After Ref. 42, with permission of
the Society of Plastics Engineers.)

own to suit the specifications. The description of several noncommer-
cial single-cell instruments of various degrees of sophistication is
available in the literature [17,28,29,32]. Some of them utilize a
photon-counting technique, provide high sensitivity, and are very
useful for conducting low-temperature nonstationary experiments.
Independent of the degree of sophistication, however, single-cell
instruments are insufficient to evaluate materials under stationary
conditions since some of such experiments may require many hours
(even days) when applied for the analysis of highly stabilized systems
and performed at relatively low temperatures.

The first commercially available multicell chemiluminograph suitable
for performing long-term stationary experiments was recently introduced by Pola Company (Figure 2a). The apparatus comprises a dark
chamber (1) with a sliding stage (2) which holds numerous individual
test cells (3) maintained on a metal support plate (4) which provides
even temperature distribution to the cells. The test cells (Figure 2b)
have a construction contributing to the warm-up of the gas before
reaching the test samples. Each cell contains metal shavings which
have a large surface area for heat exchange. The gas flow to each
sample is evenly distributed by a manifold with individual flow adjust-
ment. Each cell is covered by a glass cover to prevent cross-
contamination. Glass covers also restrict reaction volume of each cell
and promote fast replacement of one gas by another. The lower part
of the dark chamber is separated from its upper part by a metal
plate (6) with a number of holes equal to the number of test cells.
In order to prevent the heat exchange between the lower and the
upper parts of the apparatus, each hole in the separating plate is
covered by a glass window. When the sliding stage is in "in" position, each of the holes in the separating plate is strictly above one
of the test cells. The light emitted by the samples placed in the
test cells is sequentially measured by a rotating photomultiplier (7)
placed in the upper part of the dark chamber. The rotation of the
photomultiplier is provided by an electric motor (8). The light output is synchronously recorded with the sample position and assures
sequential sampling of light intensity from multisample apparatus.
In order to avoid the photomultiplier overheating during the experiments, a fan is placed in the upper part of the dark chamber. The
fan provides constant outside air circulation through the upper part
of the dark chamber. The electronic part of the apparatus (9)
consists of a photometer, a temperature programmer/controller, a
digital data-processing board, control knobs, and pilot lights.

Although the instrument described provided the opportunity of
simultaneous multisample evaluation, it possessed a deficiency in the
sense that various samples could be analyzed only at the same temperature and environmental conditions, and loading and unloading
of the cells had to be performed at the same time. This disadvantage has been overcome in the next instrument with eight completely

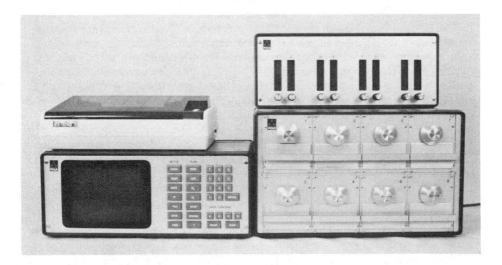

FIGURE 3   Multisample computerized chemiluminograph with eight completely independent cells.

independent cells introduced by the same company (Figure 3).   In addition to the mentioned advantage, the latter apparatus is computerized allowing fully automatic operation with the computer fulfilling two major functions: (1) control and monitoring of the temperature and atmosphere experienced by samples; (2) data storage, retrieval, and analysis.

The computer consists of two separate units: a controller and a host processor.   Each unit is controlled by its own microprocessor hardware/software system and communication between the units is carried out via a RS-232 serial interface.   The controller is an eight-channel temperature controller/programmer and a data acquisition system.   The temperature controllers are capable of independently raising the temperature at a controlled rate to a setpoint and maintaining that setpoint.   The data acquisition system inputs the individual temperature values via thermocouple sensors and light emission intensities via photomultiplier tubes.   The temperature values are fed back to the temperature control system.   They are also linearized, filtered, and sent along with the filtered emission intensity values to the host processor.   The host processor receives and stores the incoming data.   It provides channel selection as well as allows the user to enter/modify heating rates and setpoints as well as start/stop logging and temperature cycling.   It also allows the graphical representation of stored data and the results of calculations on the CRT screen as well as printing of the data.

In order to perform an experiment, a sample which may be a film, fabric, powder, or liquid is placed in a small-volume cell which

can be heated from room temperature up to 350°C at a constant heating rate varying from 1 to 15°/min. Along with the constant heating
rate mode, an isothermal mode of operation can be chosen and in
this case the desired temperature can be maintained to within ±0.2°C
in flow of either oxygen or an inert gas. The temperature of each
cell is continuously monitored with a thermocouple and displayed on
the CRT. Oxygen or the inert gas supplies are selected by two
computer-activated solenoid valves and the gas flow rate in each of
cells can be regulated by a flowmeter. The chosen gas is passed
through a heat exchanger at the bottom of the cell and then reaches
the sample. Symmetrical cell design and positions of gas inlets and
outlets assure no preferential directional flow of the gas in the cell.
Light emitted by the sample is gathered by a lens in the lid of the
cell and focused on the cathode of the photomultiplier. If desired,
the spectral distribution of the chemiluminescence can be determined
by inserting filters between the sample and the PMT. Each photomultiplier is placed into an assembly of two coaxial aluminum cylinders
having holes of the same size as the photomultiplier photocathode.
The inner cylinder can be rotated around its major axis thus opening
or closing the light pathway from the sample to the photomultiplier.
The outer cylinder is stationary and has a copper coil around it for
water circulation which provides cooling to the entire photomultiplier
housing assembly. The cooling of the PMT helps to reduce the level
of thermoelectric emission from the photocathode and is essential to
obtain a high signal-to-noise ratio at low levels of the emitted light.
The sensitivity of different photomultipliers in different cells can be
adjusted to the same level by using a standard light intensity source
and, if necessary, varying the photomultiplier voltage supply. Since
the light flux from the sample has a radial distribution, the sample-
to-photocathode distance is kept to a minimum of approximately 2 cm.
The signal output from the anode of the photomultiplier is amplified
and passed on to the computer, which functions as a data collection
terminal, simultaneously recording the time and the intensity of
emitted light. The two readings thus obtained are placed into an
array and a graphical representation of the intensity of light emitted
by the sample as a function of time is displayed on the CRT. This
information can be obtained for each of eight cells and the progress
of the experiment in each of cells can be followed simply by switching from one cell to another. After the completion of the experiment
the data stored into the computer can be used for analytical evaluation. The computer provides the opportunity for data evaluation
according to Equations (23), (26), (35), and (36) as well as calculation of the area under the curve within any chosen time interval,
and the corresponding results are displayed on the CRT. At any
time the experimental data as well as the results of calculations displayed on the CRT can be transferred to the printer to give a hard
copy of the CRT image.

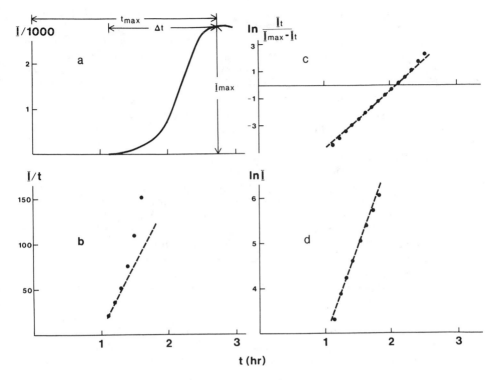

FIGURE 4   Chemiluminescence curve of unstabilized polypropylene powder obtained at 140°C (a) and the evaluation of the data according to Equations 40 (b), 23 (c), and 26 (d).

Along with the features described above, the instrument has an analog output suitable for a continuous strip chart recording.   It can also be interfaced with Apple and other computers.

VI.   INDIVIDUAL POLYMERS

A.   Polypropylene

A typical light intensity-vs.-time curve for the chemiluminescence produced by autooxidation of polypropylene consists of four regions (Figure 4a).   There is an induction period during which there is practically no light emitted by the sample, oxidation is slight, and buildup of peroxides and hydroperoxides is slow.   Following the induction period, there is an autocatalytic stage in which the hydroperoxides catalize further oxidation and the intensity of emitted light increases quite rapidly.   The induction and acceleration periods are not separate phenomena but rather parts of a typical autocatalytic reaction.   The light intensity next reaches the highest level (peak

hydroperoxide concentration). Finally, there is a period of light
decay (a deceleration of the rate of oxidation). The decrease in
oxidation rate after passing the maximum has been observed previously [30]. Possible explanations for the rate drop could involve
a decrease in the permeability of the outer surface of oxidized sample
to oxygen or the formation of reaction products which tend to inhibit
the oxidation reaction either by interaction with chain carriers or
by nonradical-induced decomposition of hydroperoxides.

For our purposes the portion of the curve 4a which extends up
to I_{max} value is of most importance since it represents the autocatalytic process. In order to establish the mechanism of oxidation
(initiation via unimolecular vs. bimolecular hydroperoxide decomposition), the data are plotted according to Equations (40), (23), and
(26) in Figure 4b to d, respectively. As one can see, the assumption of a unimolecular hydroperoxide decomposition is clearly not in
agreement with the experiment with the possible exception of the very
early stages of the reaction (Figure 4b). On the other hand, the
results can be approximated by Equation (23) very well in a broad
time interval (Figure 4c). When Equation (23) is applied, the
straight-line approximation is observed from the very early stages of
the process up to the chemiluminescence intensities approaching 75%
of the maximum value. The deviation from the straight line at
higher intensities reveals the increased complexity in the mechanism
of oxidation at its advanced stages. As is expected, Equation (26)
yields the same value of oxidation rate as Equation (23) when intensity lies between 2 and 5% of the maximum intensity (Figure 4d).
The oxidation rate values evaluated at lower (below 2%) and higher
(above 5%) values of the maximum intensity are somewhat higher and
lower, respectively. Thus the time interval where Equation (26) gives
exactly the same results as Equation (23) is quite narrow and caution
should be exercised when the former equation is used.

In the experiment just described, the unstabilized polypropylene
powder (sample A) was heated under nitrogen from room temperature
up to 140°C and then nitrogen flow switched to oxygen (zero time).
Only the second stage of the experiment is shown in Figure 4a, since
in nitrogen the original polypropylene powder did not exhibit any
observable luminescence. Nonetheless preoxidized polypropylene is
known to luminesce even under nitrogen [22]. In order to evaluate
the kinetics of decomposition of polypropylene hydroperoxides in
nitrogen, the experiment shown in Figure 4a has been extended as
follows: Upon the attainment of maximum chemiluminescence intensity
under oxygen, the heater was turned off and two identical samples
in two separate cells cooled down to 40°C when still under oxygen.
Then the gas flow was switched to nitrogen in both cells. One of
the samples was heated at the constant heating rate of 2°/min up to
220°C (Figure 5a), while the other sample was heated up to 140°C
and the glow decay studied at this temperature (Figure 5b). The
results of the nonisothermal experiment are summarized in Table 1.

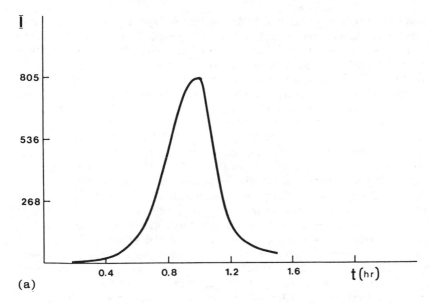

(a)

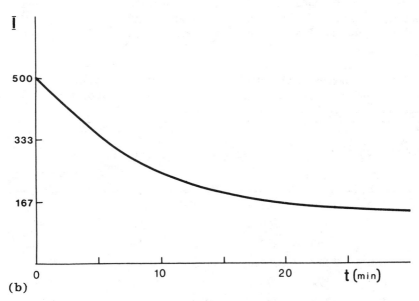

(b)

FIGURE 5   Polypropylene hydroperoxide decomposition: (a) constant heating rate (2°/min) and (b) isothermal (T = 140°C) experiment.

The reaction rates k_D and $k_D'[ROOH]_0$ for polypropylene hydroperoxide decomposition were calculated by means of Equations (48) and (54), and the values of the conversion of the reaction α were determined as the ratio of the area under the recorded curve from zero time to time t and the total area under the curve. The activation energy for polypropylene hydroperoxide decomposition was evaluated by the two independent methods: the method of initial rises and the conventional plot of ln k_D vs. 1/T (Figure 6). As can be seen, the experimental results are well approximated by straight lines and both methods give practically the same activation energy in the region from zero up to 60 to 70% of conversion (40 to 160°C temperature interval). On the other hand, the assumption of the second order decomposition led to essentially nonlinear activation energy plot for degrees of conversion exceeding 15%. In addition to these data, the first order kinetics of polypropylene hydroperoxide decomposition was also confirmed when several samples with different initial hydroperoxide concentrations were studied. These samples were obtained by the interruption of the autooxidation at early stages far before the chemiluminescence under oxygen reached the maximum value. It was found that the position of the chemiluminescence maximum under nitrogen for these samples was the same as shown in Figure 5a, namely, 160°C. The suspicion that this maximum was simply due to polypropylene melting was ruled out because the increase in the

TABLE 1   Evaluation of the Decomposition of Polypropylene Hydroperoxides from Nonisothermal Experimental Data (Unstabilized Powder A)

t(hr)	I(mV)	T(°C)	$\int_t^{1.5} I\, dt$	α(%)	$k_D(hr^{-1})$	$k_D'[ROOH]_0$ (hr^{-1})
0	0	40	302.2	0	—	—
0.4	18	88	300.5	0.6	0.06	0.06
0.5	47	100	297.3	1.6	0.16	0.16
0.6	108	112	290.0	4.0	0.37	0.39
0.7	239	124	273.2	9.6	0.87	0.97
0.8	437	136	239.9	20.6	1.82	2.29
0.9	655	148	185.2	38.7	3.54	5.77
1.0	789	160	112.0	62.9	7.04	19.01

Note: t, time; I, chemiluminescence intensity; T, temperature; α, degree of conversion; k_D, rate of the first order decomposition; k_D', rate of the second order decomposition.

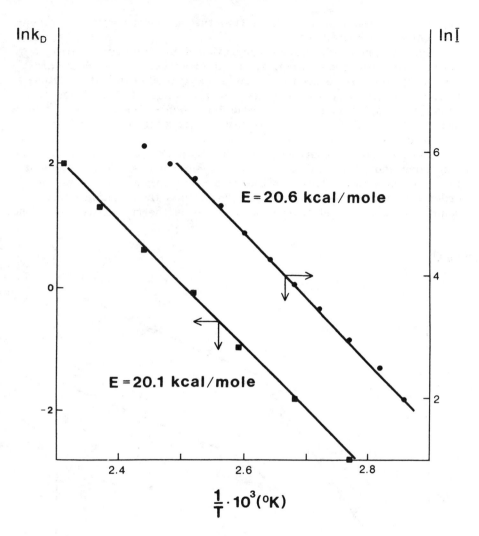

FIGURE 6   Evaluation of the activation energy of polypropylene hydroperoxide decomposition: (•) method of initial rises, (■) ln k_D vs.
1/T dependence.

heating rate from 2 to 10°/min shifted the maximum position to essentially higher temperatures.

While the nonisothermal decomposition data obtained in nitrogen suggest the first order kinetics, the indirect evaluation of polypropylene hydroperoxide decomposition in oxygen indicates the participation of two hydroperoxide molecules in a single decomposition act, thus providing an example of a reaction which is bimolecular stoichiometrically but first order kinetically. This kind of behavior is well documented in solids (60).

The isothermal experiments (140°C) with polypropylene hydroperoxides showed an initial fast decay over the first 15 min followed by a slower decay. This was the case when either first or second order kinetics was assumed (Figure 7). Thus the reaction order could not be evaluated on the basis of the isothermal decay data alone.

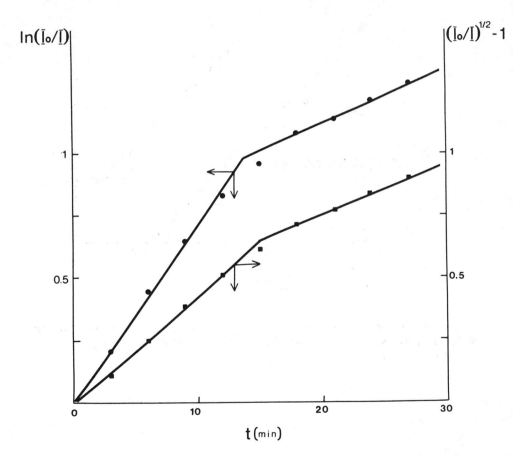

FIGURE 7 Chemiluminescence decay at 140°C: (•) first order kinetics, (■) second order kinetics.

The estimate of the rate of the first order decomposition appeared to be 4.2 and 1.2 hr^{-1} for fast and slow parts of the decay, respectively. At the same time, the nonisothermal measurements gave the intermediate value of 2.4 hr^{-1} (Table 1). These results seem to indicate that the thermal decomposition of polypropylene hydroperoxides consists of two parallel reactions which can be separated by the isothermal experiment. Similar observation was reported by Chein and Jabloner [52].

Concluding the discussion on the kinetics of polypropylene hydroperoxide decomposition it is worthwhile to mention that the activation energy values for polypropylene oxidation (19.6 kcal/mole, ref. 44) and hydroperoxide decomposition in nitrogen (Fig. 6) were found to be practically identical, indicating the validity of the basic oxidation scheme utilized.

Although polypropylene is probably the most throughly studied polymer by means of chemiluminescence, the influence of sample preparation and experimental conditions on chemiluminescence results is essentially unknown even for this material. In order to fulfill this gap, polypropylene film samples of different thicknesses were studied immediately after preparation and also after 80 days of aging under ambient conditions. Some of the chemiluminescence experiments were carried out at various rates of oxygen flow. All film samples were compression-molded from unstabilized polypropylene powder A at 190°C. In order to prevent essential oxidation, the compression molding time was kept to a minimum ($\approx$ 1 min) with following quenching in a water—ice mixture. The samples for chemiluminescence evaluation were prepared in the way similar to the sample preparation in differential scanning calorimetry where a sample is compressed in an aluminum cuvette with an aluminum plate on the top. In our case, however, each of the upper aluminum plates had a hole of 2 mm in diameter for light output and thus samples of various thicknesses were characterized by strictly the same exposed surface area. The compression of the samples between two aluminum layers was also helpful in preventing sample curling.

The sigmoidal curves were obtained for all samples studied and the corresponding data are summerized in Table 2. Several important conclusions can be made on the basis of the results obtained:

(1) Chemiluminescence emission is originated not entirely from the sample surface as is usually assumed [32]. However, the increase in the intensity of emitted light with the sample thickness is not linear and the deviation from the linearity becomes more pronounced as the sample thickness increases. This effect may be due to material self-absorption or the fact that for thick samples oxidation reaction becomes diffusion-controlled. The latter factor is believed to predominate since polypropylene is practically transparent up to the thickness of 3 mm [33].

(2) The oxygen flow rate seems to have no effect on the results. It has to be emphasized that promotion or inhibition of oxidation

TABLE 2 Chemiluminescence Data for Polypropylene Powder and Film Samples Obtained at 140°C (unstabilized sample A)

Sample	Room temp. aging (days)	Oxygen flow rate (ml/min)	Induction time ln($[A]_0/[B]_0$) (dimensionless)	Oxidation rate $k[A]_0$ (hr^{-1})	I_{max} (mV)	t_{max} (hr)	Δt (hr)
Powder	—	20	10.1	4.9	2474	2.7	1.9
Powder	—	55	9.6	4.9	3242	2.6	1.9
30 μm	0	20	4.4	7.3	310	1.1	0.9
30 μm	0	55	4.6	8.1	304	1.0	0.8
30 μm	80	20	3.8	6.3	255	1.1	0.9
30 μm	80	55	4.0	6.9	297	1.1	0.9
50 μm	0	20	5.4	8.6	531	1.1	0.9
50 μm	80	20	4.6	6.8	504	1.3	1.1
150 μm	0	20	7.5	10.1	924	1.4	1.1
150 μm	80	20	5.1	4.0	779	2.2	1.8
500 μm	0	20	7.0	9.3	1945	1.9	1.7
500 μm	0	55	6.8	9.0	2216	2.0	1.8
500 μm	80	20	4.6	2.8	2436	3.0	2.6
500 μm	80	55	3.9	2.5	2519	3.3	2.9

Note: The chemiluminescence curve parameters I_{max}, t_{max}, and Δt are shown in Figure 4a.

associated with the oxygen flow rate observed in some cases could be an artifact. If there is a temperature gradient between the sample and the gas introduced to the cell, the change in oxygen flow will lead to variations in heat exchange and as a result to a sample temperature fluctuation. Such a situation is quite probable to occur in aging ovens with enforced air circulation. Nonetheless the conclusion regarding the independence of the chemiluminescence results on the oxygen flow rate may not be of a common nature. In some cases the effect of gas flow on additive (or reaction product) volatilization may be of prevailing importance. Thus the effect of the gas flow on the oxidation kinetics has to be evaluated for each particular system.

(3) The induction time values for all compression-molded samples are essentially lower than those for the original powder. This shows that, despite all precautions some oxidation did take place in the process of sample preparation. Thin films seem to be influenced by compression molding to a larger extent than thick ones. Also, the induction time decreases slightly during aging, the effect being more pronounced in thick samples.

(4) The oxidation rate for original powder lies within the range of oxidation rates obtained for different film samples, although all freshly prepared samples have higher oxidation rates than original powder. Upon aging, there is a clear tendency toward the decrease in the rate of oxidation, especially for thick samples.

(5) As a result of aging, neither a shift of a chemiluminescence curve as a whole nor its spread out along the time scale was noticed for thin samples. Contrary to this, an essential spread out of the chemiluminescence curve toward longer times was noticed for freshly prepared and especially for aged thick samples. This indicates that oxidation is diffusion-controlled in thick films and that certain structural changes during aging promote this effect.

Basic changes in the parameters of oxidation of aged samples observed by us, i.e., the constancy of the induction time and the decrease of the oxidation rate values with the sample thickness for films thicker than 30 μm are the same as reported by Boss and Chien [34]. The effect of room temperature aging on structural rearrangements and in particular on the relaxation transitions in polypropylene has been repeatedly noted previously. A gradual increase in T_g of isotactic polypropylene during storage at room temperature has been observed and examined in detail [35,36]. The latter effect is usually explained by the fact that the amorphous chains are fairly rigidly fixed and are subjected to high constraint by surrounding crystallites. Gradual structural ordering during secondary crystallization forces the amorphous chains into a more strained state, so T_g is increased. The consequence of this effect should be an increase in the degree of rigidity of the polymer and a decrease in the rate of oxygen diffusion into the material and as a result a decrease in oxidation rate

with aging time. The extent of the decrease in oxidation rate with
aging can be expected to be larger in thick films and this is confirmed
experimentally.

Along with long-term room temperature aging, the high-temperature
annealing under nitrogen also brings about changes in polypropylene
oxidation parameters [37]. For a material exposed to high temperatures under an inert atmosphere, it can be visualized that thermally
initiated hydroperoxide decomposition (initiation step) will not be
followed by propagation reactions which require the presence of oxygen. Instead, some of the radicals formed in the initiation step may
undergo isomerization, which leads to induced hydroperoxide decay
[38]. In low molecular weight analogs of polypropylene, isomerization
of the radicals in the liquid phase is appreciable at oxygen pressures
below 200 torr [39]. In the absence of the propagation step, radicals
may also recombine either by diffusively approaching each other or
by a chain transfer mechanism, although these processes seem to be
less probable in the solid state than intramolecular radical isomerization. Each of the above reactions, if it takes place, will lead to a
decrease in the hydroperoxide concentration. On subsequent oxidation, the decrease in the initial effective hydroperoxide concentration
should be accompanied by an increase in induction time. Indeed, this

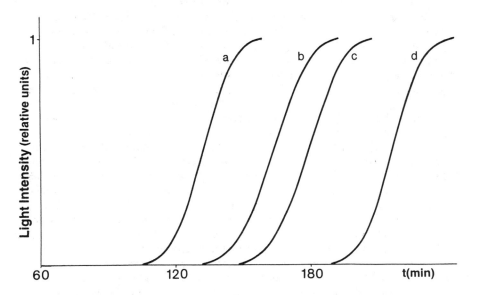

FIGURE 8   Chemiluminescence curves of polypropylene samples annealed at 150°C under nitrogen atmosphere for various times: (a) 0,
(b) 2.6, (c) 5, (d) 11 hr. (After Ref. 37, with permission of John
Wiley and Sons, Inc.)

TABLE 3 Parameters of Autooxidation at 150°C for Polypropylene Samples Annealed at This Temperature Under Nitrogen Atmosphere for Various Times

Annealing time (hr)	Induction time $\ln([A]_0/[B]_0)$ (dimensionless)	Oxidation rate $k[A]_0$ (hr^{-1})
0	14.1	6.4
2.6	15.5	5.7
5.0	17.9	6.0
11.0	23.4	6.5

Source: Ref. 37.

is the type of behavior observed in both stabilized and unstabilized polypropylene. The chemiluminescence curves obtained for slightly stabilized polypropylene 40-mesh powder samples annealed under nitrogen atmosphere at 150°C for various times prior to introduction of oxygen are shown in Figure 8. Table 3 presents the induction time and oxidation rate values calculated according to Equation (23). As can be seen, annealing of the material results in a steady increase in the induction time, leaving the oxidation rate practically unchanged.

In order to estimate the activation energy of polypropylene autooxidation in solid state, unstabilized polypropylene powder B was studied at five different temperatures (Figure 9). The results obtained ($\ln([A]_0/[B]_0)$ and $k[A]_0$ values) and also the maximum light intensities are shown in Table 4. It has to be underlined that the area under each of the curves in Figure 9 was found to be about the same, indicating that the same amount of hydroperoxide is formed at various temperatures before the oxidation chain length reaches unity. Two different approaches have been used for activation energy evaluation: (1) The conventional plot of $\ln(k[A]_0)$ vs. $1/T$ and (2) The method originally developed for isothermal solid-state decomposition reactions studied by DTA [40], where it was suggested that the slope of a plot of $\ln(H_{max})$ vs. $1/T$ yields the activation energy (H_{max} is the maximum peak height of the isothermal DTA trace). The activation energies in the 110 to 150°C temperature interval were 19.4 and 23.3 kcal/mol, respectively, according to the two methods (Figure 10a). Although the second value coincides precisely with the activation energy of polypropylene oxyluminescence reported by Schard and Russell [10] whereas the first value is slightly lower, it is believed that the direct method of determining the activation energy provided by the logarithm of the reaction rate vs. reciprocal temperature is the more reliable.

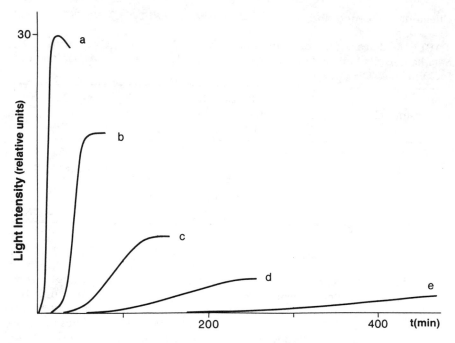

FIGURE 9 Chemiluminescence curves of polypropylene at different temperatures: (a) 150, (b) 140, (c) 130, (d) 120, (e) 110°C. (After Ref. 44, with permission of John Wiley and Sons, Inc.)

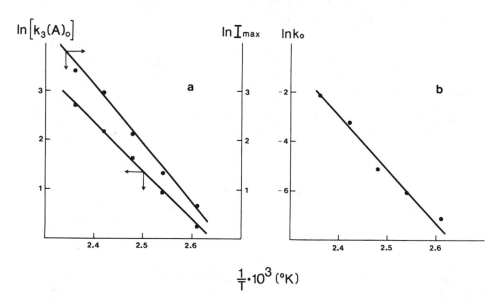

FIGURE 10 Activation energy of polypropylene autooxidation: (a) over-all oxidation, (b) initiation step.

TABLE 4 Autooxidation Parameters for Unstabilized Polypropylene Powder B

Temperature (°C)	Induction time $\ln([A]_0/[B]_0)$ (dimensionless)	Oxidation rate $k[A]_0$ (hr^{-1})	Initiation rate k_0 (hr^{-1})	I_{max} (rel. units)
150	4.7	12.7	0.12	30
140	5.4	9.0	0.042	18.9
130	6.7	4.6	0.0059	8.2
120	7.0	2.5	0.0024	3.7
110	7.4	1.3	0.00083	1.9

Source: Ref. 42.

   The induction time $\ln([A]_0/[B]_0)$ values presented in Table 4
were obtained with the assumption of the presence of hydroperoxide
seeding at the beginning of oxidation.   If this assumption is omitted,
and instead two parallel reactions (namely a noncatalytic first order
transformation and an autocatalytic transformation of A into B) are
considered, the rate of initiation k_0 can be calculated according to
Equation (29).   Corresponding results are also shown in Table 4.
As it follows from Table 4, the rates of initiation are essentially
lower than rates of propagation, especially at low temperatures.   The
evaluation of the activation energy of initiation (Figure 10b) gives
the value of 43 kcal/mol.   Although this value correlates quite well
with the Uri's estimate of the activation energy required for direct
interaction between a hydrocarbon and oxygen [43], it is much larger
than the activation energy for polypropylene autooxidation and such
a reaction seems not to commend itself from a physicochemical point
of view.   Thus polypropylene oxidation most probably proceeds via
decomposition of hydroperoxides initially present in the sample rather
than by thermally initiated hydroperoxide formation.
   The major purpose of our work was to develop a chemiluminescence
method for predicting the long-term stability of materials.   Table 5
shows the results of comparative evaluation of four polypropylene
samples by the conventional oven-aging test and the chemiluminescence technique.   Compression-molded plaques and powder samples
were used in the former and the latter cases, respectively.   One
can conclude that both techniques give correlative results: long oven
lives correspond to long induction times and low oxidation rates.   It

TABLE 5   Evaluation of Polypropylene Thermal Oxidative Stability
by the Chemiluminescence and Oven-Aging Methods

| Sample | Chemiluminescence analysis (150°C) | | Oven life (150°C) (days) |
	Induction time $\ln([A]_0/[B]_0)$ (dimensionless)	Oxidation rate $k[A]_0$ (hr^{-1})	
C	8.3	6.0	2
D	53.6	0.7	106
E	14.5	2.4	12
F	22.8	2.4	74

Source: Ref. 42.

has to be emphasized that the time required for the chemiluminescence analysis was 2 hr for sample C and 25 hr for sample D, compared to 48 hr and 106 days, respectively, in the case of the oven-aging test. The other important advantage of the chemiluminescence technique is the possibility of obtaining the quantitative information (induction time and oxidation rate), whereas the failure point of the oven-aging test is defined as the first observation of powdery disintegration or brittleness, and thus is essentially qualitative.

B. Polyethylene

The application of chemiluminescence to the study of polyolefins has been almost exclusively restricted to polypropylene. Polyethylene attracted much less attention in particular because of the low emission intensity and thus the necessity to employ sensitive instrumentation.

In our studies of a variety of polyethylenes (low-, linear low-, and high-density samples) an essential peculiarity in the chemiluminescence response was noticed, namely, instead of a single sigmoidal curve two to some extent overlapping S-shape curves were registered (Figure 11). Although Figure 11 shows only the data for low-density

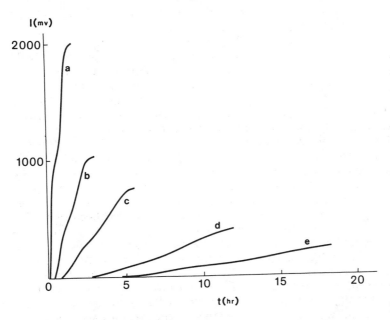

FIGURE 11   Chemiluminescence curves for low-density polyethylene at different temperatures: (a) 190, (b) 180, (c) 170, (d) 160, (e) 150°C.

polyethylene, similar results were obtained with linear low- and high-density polyethylenes.   Thus it seems polyethylene oxidation consists of two consecutive autooxidation processes which we are going to consider the first and second stage oxidations.   In this respect it is of interest to mention that two clearly resolvable processes, fast and slow, were noticed by J. C. W. Chien [57] during polyethylene hydroperoxide decomposition.   If one bears in mind that polyethylene oxidation is initiated by the decomposition of hydroperoxides, the close relationship between Chien's and our results becomes apparent.

Although the complicated shape of the curves in Figure 11 makes the application of the Equation (23) somewhat ambiguous, both first and second stage oxidations were evaluated for each of the curves in Figure 11 and the results are summerized in Table 6.   The oxidation rate values show steady increase with temperature for both first and second stage processes.   At the same time, essential scattering in the induction time can be noticed and it seems that the latter is temperature-independent and just fluctuates around a certain average value.   This is understandable since the induction time only depends on the initial hydroperoxide/hydrocarbon ratio.

The activation energy of oxidation (Figure 12) is the same (26 kcal/mol) for both first and second stage oxidation processes and practically identical with the activation energy for fast and slow polyethylene hydroperoxide decomposition evaluated by Chien (25 and 27 kcal/mol, respectively) [57].

The kinetics of polyethylene hydroperoxide decomposition in nitrogen has been evaluated in the manner similar to that described previously for polypropylene.   Two identical polyethylene samples in two separate cells were oxidized at 190°C for different lengths of time.   One of the samples (sample A) was kept under oxygen until the completion of both oxidation stages, whereas the other (sample B) was oxidized only for the time sufficient to complete the first stage.   Then each of the samples was cooled down to 60°C when still under oxygen, the gas flow switched to nitrogen in both cells, and the samples heated at the rate of 1°/min up to 200°C.   The results of the nonisothermal experiments under nitrogen are shown in Figure 13.   The lower total intensity of emitted light from sample A as compared to sample B is most probably due to partial decomposition of hydroperoxides accumulated during the first stage of oxidation.   Also the hyroperoxide decomposition curve for sample A which undergone both stages of oxidation is clearly nonsymmetrical indicating more than one decomposition reaction.   Thus only curve B was utilized for the evaluation of the parameters of decomposition.   The values of k_D and $k_D'[ROOH]_0$, the first and second order hydroperoxide decomposition rates, as well as α, the degree of conversion, are shown in Table 7.   The data in parentheses are the first order rate constants obtained by Chien [57] for fast decomposition.

TABLE 6 Autooxidation Parameters for Low-Density Polyethylene

Temperature (°C)	First stage		Second stage	
	Induction time $\ln([A]_0/[B]_0)$ (dimensionless)	Oxidation rate $k[A]_0$ (hr^{-1})	Induction time $\ln([A]_0/[B]_0)$ (dimensionless)	Oxidation rate $k[A]_0$ (hr^{-1})
150	6.1	0.8	9.5	0.6
160	7.6	1.5	9.3	1.0
170	7.8	4.4	8.4	2.1
180	4.5	5.9	8.5	4.0
190	5.3	10.1	9.3	9.0

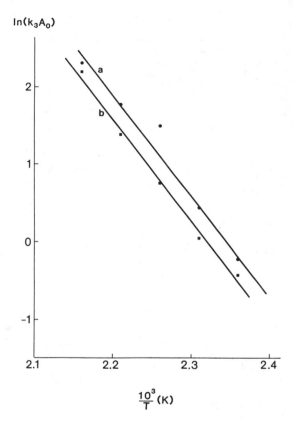

FIGURE 12   Evaluation of the activation energy of oxidation for low-density polyethylene: (a) the first stage, (b) the second stage.

The activation energy plot of hydroperoxide decomposition is well approximated by a straight line in the entire region of conversion studied if the first order reaction is assumed, whereas the attempt to represent the data according to the second order decomposition failed (Figure 14). Surprisingly, the activation energy of decomposition in nitrogen was found to be about 15 kcal/mol, which is essentially lower than the 25 kcal/mol value reported by Chien [57]. Considering the differences in polyethylene preparation, it can be assumed that in our case reaction is catalyzed by products of high-temperature oxidation.

In order to demonstrate the applicability of chemiluminescence for comparative evaluations, the study of unstabilized and stabilized high-density polyethylene samples at different temperatures was conducted and the corresponding results are shown in Figure 15. Again

TABLE 7 Evaluation of the Decomposition Parameters for Polyethylene Hydroperoxides

t(hr)	I(mV)	T(°C)	$2.3 \int_t Idt$	$\alpha(\%)$	$k_D(hr^{-1})$	$k'_D[ROOH]_0$ (hr^{-1})
0	0	60	87.4	0	—	—
0.6	13	96	84.6	3.2	0.15	0.16
0.67	14	100	83.6	4.3	0.17 (0.11)[a]	0.17
0.8	17	108	81.7	6.5	0.21	0.22
1.0	31	120	77.0	11.9	0.40 (0.61)[a]	0.46
1.2	49	132	69.3	20.7	0.71	0.89
1.25	56	135	65.6	24.9	0.85 (2.25)[a]	1.14
1.4	73	144	57.4	34.3	1.27	1.94
1.6	85	156	41.6	52.4	2.04	4.30
1.8	87	168	24.1	72.4	3.61	13.07

[a]Ref. 57.

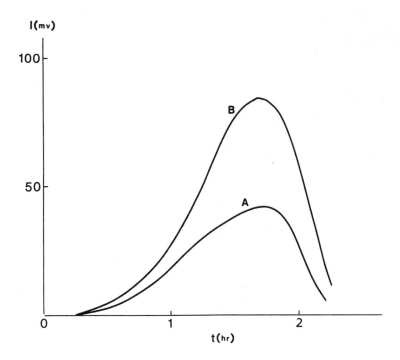

FIGURE 13 Polyethylene hydroperoxide decomposition under nitrogen atmosphere. Heating rate 1°/min.

two consecutive S-shape curves characterize each of the samples. The induction time and oxidation rate values calculated for the first and second stage oxidation (Table 8) clearly reveal the superior quality of the stabilized material.

C. Acrylonitrile-Butadiene-Styrene Copolymers

Similarly to polyolefins, acrylonitrile-butadiene-styrene (ABS) copolymers, if not purposly preoxidized, practically do not emit light under the nitrogen atmosphere. The most essential difference in the character of chemiluminescence vs. time curves between polyolefins and ABS is the initial burst of emission observed for ABS samples when the atmosphere is changed from nitrogen to oxygen. The initial increase in the light intensity immediately after the introduction of oxygen has been observed previously and attributed to the presence of easily oxidizable centers in a polymer at the beginning of the chemiluminescence experiment when accumulated hydroperoxides are not as yet present [21].

TABLE 8   Autooxidation Parameters for Stabilized and Unstabilized High-Density Polyethylene Samples

Temperature (°C)	First stage		Second stage	
	Induction time $\ln([A]_0/[B]_0)$ (dimensionless)	Oxidation rate $k[A]_0$ (hr^{-1})	Induction time $\ln([A]_0/[B]_0)$ (dimensionless)	Oxidation rate $k[A]_0$ (hr^{-1})
150	3.9 (9.3)	1.0 (0.7)	10.2 (14.1)	1.1 (0.7)
190	2.3 (4.9)	8.6 (6.8)	5.5 (7.7)	4.3 (4.1)

Note: Data in parentheses are given for the stabilized sample.

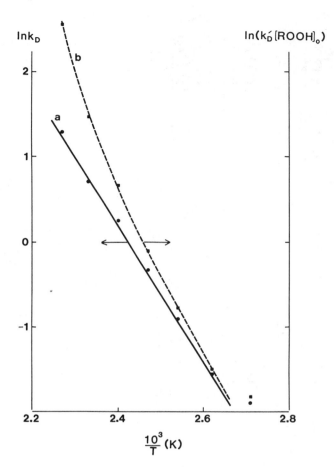

FIGURE 14   Evaluation of the activation energy of polyethylene hydroperoxide decomposition in nitrogen: (•) first order decomposition,
(■) second order decomposition.

The experimental chemiluminescence results obtained for two ABS
powder samples at 150 and 190°C are presented in Figure 16.   The
chemiluminescence experiments at 190°C have been performed in order
to be able to compare the data with the DSC and oxygen uptake
results since the sensitivity of the two latter techniques is not sufficient for the application at 150°C.

Besides exhibiting the initial burst of emission in oxygen, the
other difference between polypropylene and ABS is that the latter
seems to exhibit a two-stage oxidation, which, however, is different
from the two-stage polyethylene oxidation.   With ABS, the first stage

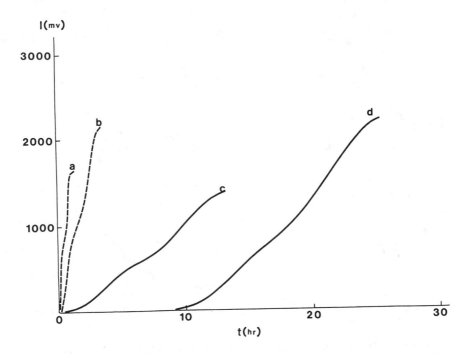

FIGURE 15   Chemiluminescence curves for high-density polyethylene samples: (a,c) unstabilized, (b,d) stabilized samples; (a,b) 190, (c,d) 150°C.   The intensity of emitted light for samples c and d is magnified 10 times.

shows a slow, steady increase in the chemiluminescence intensity, whereas the second stage is characterized by a sigmoidal curve typical for an autooxidation process (Figure 16).   A similar two-stage degradation was recently reported for styrene-isoprene-styrene block copolymers when the initial steady rate of chain scission and steady buildup of oxidation products accompanied by no change in surface tension was followed by a rapid reduction in molecular weight through an oxidation chain-scission mechanism accompanied by a rapid increase in surface tension [45].

The chemiluminescence data together with the DSC and oxygen uptake results are shown in Table 9.   Since the chemiluminescence experiment at 190°C was completed within several minutes, the kinetic approach according to Equation (23) was not used.   Instead the time to reach the maximum intensity (t_{max}) and the ratio of the initial burst of emission (I_{in}) to the maximum intensity (I_{max}) were measured.   As it follows from Table 9, there is basically a good

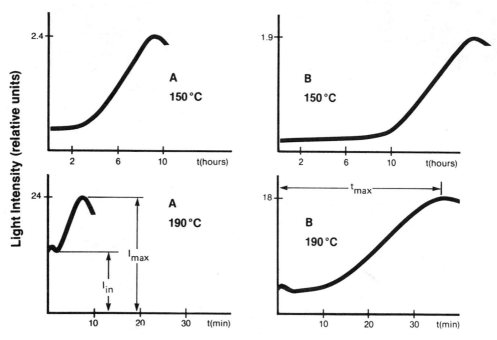

FIGURE 16 Chemiluminescence curves of two acrylonitrile-butadiene-styrene copolymer samples A and B obtained at 150 and 190°C. (After Ref. 42, with permission of the Society of Plastics Engineers.)

correlation between the induction time values obtained by the DSC and oxygen uptake methods and the time to reach the maximum intensity (the chemiluminescence method). Only sample A was the exception. The induction time for this sample was evaluated as "very short" by the DSC and oxygen uptake techniques and was smaller than for the other samples. On the other hand, t_{max} was similar for the samples A, C, D, and E when estimated by the chemiluminescence method. There is a better correlation between the DSC, oxygen uptake, and chemiluminescence results when instead of t_{max} the value I_{in}/I_{max} is used. At a first approximation the I_{in}/I_{max} ratio is indicative of the amount of easily oxidizable sites on polymer surface. Thus, it seems that very short induction time obtained for sample A in this particular case is really not the induction time of an autocatalytic process but rather is associated with the essential content of unstable products which, however, do not autocatalyze oxidation. Further indication on that was obtained by the chemiluminescence experiment performed at 150°C (Table 9). At this temperature sample A exhibited longer induction time and smaller oxidation rate than samples C, D, and E, although sample B remained the most

TABLE 9 Evaluation of ABS Thermal Oxidative Stability by the Chemiluminescence, DSC, and Oxygen Uptake Methods

Sample	DSC 190°C Induction time (min)	Oxygen uptake 190°C Induction time (min)	Chemiluminescence 150°C Induction time $\ln([A]_0/[B]_0)$ (dimensionless)	Chemiluminescence 150°C Oxidation rate $k[A]_0$ (hr^{-1})	Chemiluminescence 190°C t_{max} (min)	Chemiluminescence 190°C I_{in}/I_{max}
A	V.short	V.short	5.3	0.84	7	0.46
B	34	43	10.7	0.72	35	0.17
C	13	11	3.8	1.02	9	0.24
D	11	5	3.8	1.20	6	0.26
E	13	10	4.2	1.20	8	0.21

Source: Ref. 42.

stable. It should be also emphasized that, as indicated previously, the induction and acceleration periods are not separate phenomena but parts of a typical autocatalytic process. Thus both these parameters should be considered together and the utilization of the induction time only by the DSC and oxygen uptake methods may not be appropriate.

As it is known, ABS exhibits a dramatic loss of impact resistance when aged in the air oven, even at 130°C. The DSC and oxygen uptake methods are probably useful in estimating the high-temperature performance but not necessarily directly applicable to predict lifetimes at service temperatures [46]. Thus the ability of chemiluminescence to be applied for evaluation of ABS thermal oxidative stability at 150°C and even lower temperatures seems to be important.

So far the results of evaluation of ABS powders have been presented. Now we are going to turn our attention to the chemiluminescence data obtained with ABS film samples. The ABS sample F was used for the preparation of compression-molded specimens and the procedure of film sample preparation was the same as one described above for polypropylene.

First of all, it has to be emphasized that differently from polypropylene films all ABS film samples exhibited some light when heated under nitrogen, which revealed greater ABS suspectability to oxidation in the process of sample preparation. The level of the emitted light, however, was independent of sample thickness indicating the formation of hydroperoxides mostly on the sample surface. A subsequent switch to oxygen at 150°C was accompanied by a burst of emission, which was also the same for films of different thicknesses providing an additional proof that oxidation associated with compression molding was mainly restricted to the sample surface.

The samples analyzed just after preparation showed very rapid oxidation where the initial burst of emission and the following autocatalytic oxidation were essentially overlapped making questionable any quantitative analysis of the results (Figure 17a). A room temperature aging, however, was accompanied by a clear separation of these two stages and the corresponding data obtained for samples of different thicknesses aged at ambient conditions for 80 days are shown in Table 10. The results presented in Table 10 and discussed previously for polypropylene film samples (Table 2) are essentially similar and thus the major conclusions reached for polypropylene seem to be valid for ABS as well. There was, however, one basic dissimilarity in the results: the maximum intensity of the light emitted by the ABS samples decreased as the film thickness increased (Figure 17, Table 10). We were essentially puzzled by this observation until results of a similar nature were found in the literature. In the study performed by Harrison et al. [45] it was shown that a surface of styrene-isoprene-styrene block copolymer films oxidized slower than the bulk and the following explanation originally offered by Kaelble

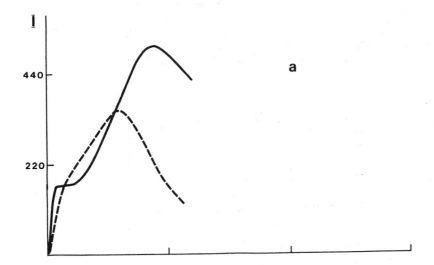

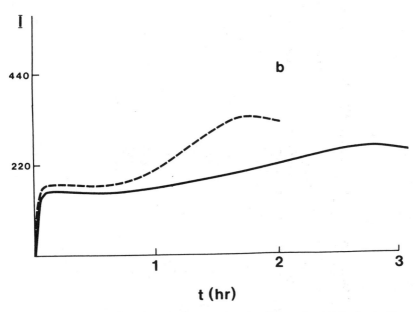

FIGURE 17 Chemiluminescence curves of acrylonitrile-butadiene-styrene copolymer film samples: (a) 30-μm film, aging time 0 (---) and 80 days (——); (b) 75-μm (---) and 220-μm (——) films, aging time 80 days.

TABLE 10 Chemiluminescence Data for ABS Powder and Film Samples Obtained at 150°C (Sample F)

Sample	Room temp. aging (days)	Oxygen flow (ml/min)	Induction time $\ln([A]_0/[B]_0)$ (dimensionless)	Oxidation rate $k[A]_0$ (hr^{-1})	I_{max} (mV)	t_{max} (hr)	Δt (hr)
Powder	—	20	15.6	6.9	2160	1.5	0.8
Powder	—	55	14.9	7.0	1944	1.4	0.7
30 μm	80	20	6.7	9.5	558	0.8	0.7
30 μm	80	55	5.3	9.3	549	0.9	0.7
75 μm	80	20	4.6	4.2	388	1.7	1.2
75 μm	80	55	4.9	3.8	417	1.8	1.4
220 μm	80	20	4.4	2.6	279	2.7	2.2
220 μm	80	55	4.1	2.8	254	2.9	2.4

[47] was given: during oxidation, polar—nonpolar molecules located
on the surface undergo a spontaneous reorientation to minimize the
surface free energy and as a result the oxidized parts of surface
molecules migrate to the bulk while lower surface energy, unoxidized
parts replace them. In chemiluminescence terms it would mean that
after initiation of oxidation on the surface slightly oxidized material
migrates from the outer transparent part of the sample into the inner,
less transparent part where its oxidation becomes controlled by the
rate of oxygen diffusion and cannot proceed as fast as on the surface.
Thus the chemiluminescence curve is expected to be spread out
toward longer times, the effect being more pronounced for thick
samples. Also, since the near-surface layers contribute most to the
intensity of emitted light and these layers are continuously replaced
by not oxidized or less oxidized, i.e., less emitting, material, the
lower intensity of emitted light from thick less transparent samples
is also understandable.

The effects described above are not to be expected in very thin
transparent films with no kinetic limitations as due to the rate of
oxygen diffusion. Since Δt values were found to be the same for
ABS powder and 30-μm film (Table 10), it can be speculated that the
latter fulfills this requirement.

D. Nylons

In contrast to polypropylene, polyethylene, and ABS, nylon 6 emits
weak light even when heated in nitrogen. Both unstabilized and stabilized powder samples exhibit very weak, practically constant light emission in the temperature interval 40 to 150°C (Figure 18). At higher
temperatures the light intensity increases exponentially and when the
melting point is reached there is a sharp decrease in the light emission,
most probably because of a sudden reduction in the surface area.

Further experiments with both unstabilized and stabilized samples
showed that any kind of heat treatment is accompanied by a rise in
the emitted light intensity especially at low temperatures. Furthermore, the chemiluminescence response depends on the time the
material is kept at ambient conditions after aging prior to the analysis
(Figure 19). The intensity of the emitted light in the low-temperature region increases with the time of exposure to ambient conditions,
whereas the high-temperature part of the chemiluminescence curve
remains practically unchanged (Figure 20). The low-temperature
peak occurs only on first heating; if polymer is cooled under nitrogen
and then reheated, the first peak does not recur, although the high-
temperature part of the curve remains unaffected and is reproduced.

Additional results, as related to the low-temperature maximum
observed under nitrogen, were published by Billingham and O'Keefe
[48] for nylon 6,6. They reported that the shape and intensity of
the peak were not affected when the cell was flushed with nitrogen

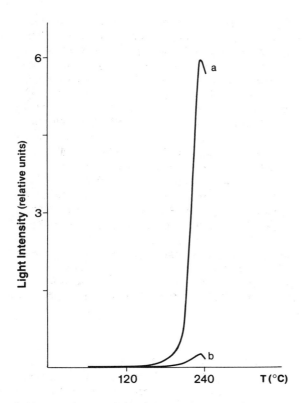

FIGURE 18 Chemiluminescence curves of unstabilized (a) and stabilized (b) nylon 6 samples. N_2 atmosphere, constant heating rate. (After Ref. 44, with permission of John Wiley and Sons, Inc.)

for 30 min or evacuated to 10^{-3} torr for 24 hr prior to heating. The other important observation was that the area of the peak can be reduced by 75% if the sample is exposed to SO_2 for 12 hr before heating. The latter treatment is known to destroy hydroperoxides by converting them to alkyl sulfates [49]. Contrary to our results, however, it was concluded that the area of the low-temperature peak decreases upon aging of the sample at 80°C in air. We believe that this controversy could arise because the samples were not exposed (or exposed for too brief a time) to ambient conditions prior to the analysis. As it follows from Figure 20, the time of about 60 hr may be needed to reach equilibrium. It was originally assumed that the long time is due to the fact that solubility and chemisorption of oxygen in a polymer decrease with increasing temperature and thus it takes some time to reach the equilibrium level of chemisorbed oxygen at room temperature [50]. However, we now believe that it is

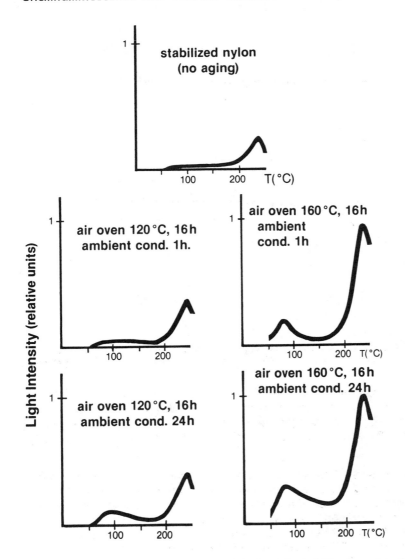

FIGURE 19 Chemiluminescence curves of stabilized nylon 6 sample aged at 120 and 160°C for 16 hr and then exposed to ambient conditions for various times. N$_2$ atmosphere, constant heating rate. (After Ref. 42), with permission of the Society of Plastics Engineers.)

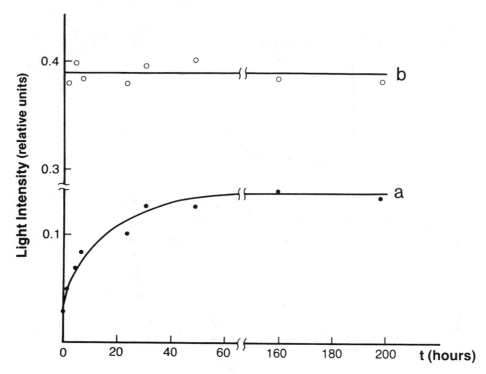

FIGURE 20   Dependence of the intensity of emitted light at 80°C (a)
and 230°C (b) vs. the time of exposure to ambient conditions for
stabilized nylon 6 sample aged in the air oven at 120°C for 16 hr.
(After Ref. 42, with permission of the Society of Plastics Engineers.)

not necessary to invoke the chemisorption concept in order to explain
the data.   Instead, it can be assumed that the restoration of the
hydroperoxides thermally decomposed in nitrogen is controlled by the
rate of oxygen diffusion into a polymer at ambient conditions.

The results described above show conclusively that chemiluminescence in 40 to 150 and 160 to 230°C temperature intervals has
different origin.   The latter conclusion is not surprising since it is
known that nylon possesses at least two major types of emitting
species (see Chapter 2).   The low-temperature emission can be confidently attributed to the decomposition of hydroperoxides, and it
was shown that the spectral distribution of the emission from nylon
6,6 oxidizing at 125°C is similar to the phosphorescence spectrum of
oxidized nylon, which was attributed to emission from the first excited
triplet of an α, β-unsaturated carbonyl group [51].   However, the
mechanism of the reaction is not clear.   The activation energy of 37
kcal/mol obtained by us [42] indicated unimolecular hydroperoxide

decomposition. Billingham and O'Keefe [48] claimed an almost four times lower value ($\approx$10 kcal/mol) and reported that the reaction follows second order kinetics. Since such a low activation energy seems too low even for bimolecular hydroperoxide decomposition, it was stated that although the initial luminescence is caused by the decomposition of hydroperoxides, it results from recombination of the product radicals. Although more research is needed to establish the precise mechanism of the low-temperature chemiluminescence, it is clearly associated with the history and processing of the sample and with the natural aging of the polymer, and the area under the low-temperature chemiluminescence peak can be used as a monitor of material quality [44].

So far we have discussed the chemiluminescence results obtained under nitrogen atmosphere and indicated that there are two distinct temperature regions with the different character of luminescence. This is also the case for the nylon luminescence under oxygen. A typical isothermal (125°C) experiment for stabilized and extracted nylon 6,6 fibers was discussed in the preceding chapter (Figure 1 in Chapter 3). The familiar sigmoidal curves are observed and can be evaluated by utilizing the approach developed for the initial stages of oxidation [Equation (23)]. At temperatures exceeding 150°C, however, nylon oxidizes without a noticeable induction period. Figure 21 presents the results obtained for unstabilized and stabilized nylon 6 at 150, 170, and 190°C. All samples exhibit a burst of emission when oxygen is introduced into the system (zero time). Then the light emission from unstabilized nylon increases at 150 and 170°C and decreases at 190°C. The equilibrium level of chemiluminescence for stabilized nylon is very quickly reached at 150 and 170°C, whereas at 190°C, similarly to the case with unstabilized material, there is a decay in light emission. When the growth and decay to the steady state is plotted according to Equations (35) and (36), a poor fit is observed for unstabilized nylon at 190°C. This is shown in Figure 22. The failure to express the experimental results by Equations (35) and (36) is not surprising since they were deduced with the assumption that hydroperoxide decomposition is the principal source of light. The fact that the glow above 150°C can be observed on repeated heatings under nitrogen indicates that this may not be the case. The attempt to express the data according to the equation originally developed for alkyl peroxy radicals recombination in liquid hydrocarbons (61) and later adapted by George for solid polymers (Equation (15) in Chapter 3) also failed.

Even when chemiluminescence is related to hydroperoxide decomposition, several problems in evaluating advanced stages of oxidation are to be expected:

1. Comparative evaluation of different samples can be made only if they exhibit relatively prolonged growth or decay of light emission

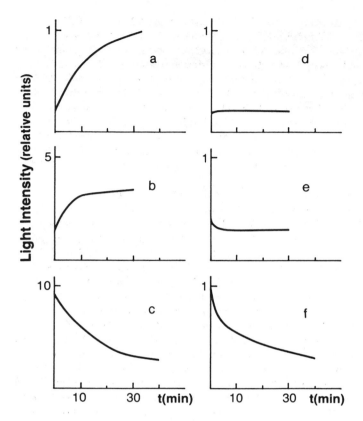

FIGURE 21   The growth and decay of light emission for unstabilized
(a, b, c) and stabilized (d, e, f) nylon 6 samples at 150 (a, d),
170 (b, e), and 190°C (c, f) after heating in nitrogen and then admitting oxygen at zero time.   (After Ref. 44, with permission of John
Wiley and Sons, Inc.)

before reaching the equilibrium (e.g., unstabilized and stabilized
nylons cannot be compared at 150 and 170°C because the equilibrium for the stabilized sample is reached at these temperatures
almost instantly).

2.   In some cases it is difficult to establish a reliable equilibrium
level of light emission since the light growth or decay may continue over a long period of time.   This might be a serious
obstacle because the evaluation according to Equations (35) and
(36) is sensitive to I_{max} and I_{min} values.

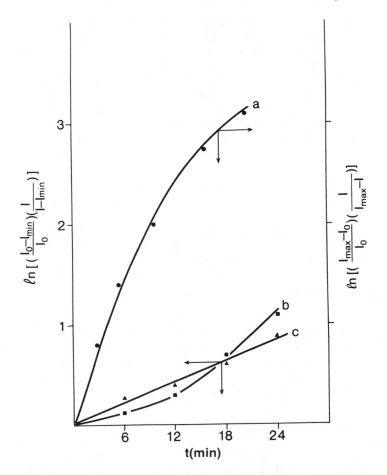

FIGURE 22 Analysis of the chemiluminescence emission growth and
decay according to Equations (35) and (36): (a) unstabilized nylon
(150°C); (b) unstabilized nylon (190°C); (c) stabilized nylon (190°C).
(After Ref. 44, with permission of John Wiley and Sons, Inc.)

    In addition, advanced stages of oxidation are complicated by
processes not taken into account. Some of the complications are as
follows: (1) the products of oxidation may play an important role as
inhibitors or activators; (2) impurities may be very important as
inhibitors or activators; and (3) a side reaction or the chain carry-
ing radicals themselves may cause destruction of hydroperoxides.

Thus it can be concluded that the application of chemiluminescence
for the evaluation of advanced stages of oxidation is restricted by
both the ability of the technique and the complex nature of the process.   It is doubtful, however, that the rate constants measured in
the region where chain lengths are close to unity and the hydroperoxide concentration is reaching its limiting value are of significance
[13].   The essential changes reflected in a material's mechanical
properties occur during the induction and autoacceleration periods
and the evaluation of the useful lifetime of a polymer should be based
on the determination of the parameters of oxidation in these two
regions.

VII. CONCLUSIONS

The first publications concerning the possibility of using chemiluminescence as the method for evaluating polymer thermal oxidative
stability appeared about 25 years ago and since then many separate
findings in this field have been published.   Nonetheless, for a long
time the technique remained a matter of discovering conditions under
which the light emission could be related to the properties of interest.
Only recently has the emphasis of the research shifted to the development of quantitative methods of the analysis.   Although there are
still many unsolved problems and further detailed studies of different
polymeric systems are required, chemiluminescence already has been
proven to be one of a few techniques capable of providing important
kinetic information related to polymer long-term stability. Graphs.

APPENDIX: CHEMILUMINESCENCE IN EVALUATION
OF INHIBITED OXIDATION

The sequence of the elementary reactions (1) through (6) is applicable to uninhibited autooxidations.   If an inhibitor InH is added to
the system, then at least two further elementary reactions should be
considered:

$$RO_2^{\cdot} + InH \xrightarrow{\ k_7\ } ROOH + In^{\cdot} \qquad\qquad (62)$$

$$RO_2^{\cdot} + In^{\cdot} \xrightarrow{\ k_8\ } ROOIn \qquad\qquad\qquad (63)$$

In the presence of the sufficient quantity of inhibitor, the normal
chain termination reactions (4) through (6) would be completely replaced by the reactions (62, 63).   Reaction (63) which is the recombination of the radicals, does not require any activation energy except
for the kinetic energy needed for the translational motion.   Conse-

quently, the k_8 value is very large and the rates of the consecutive reactions (62) and (63) are determined by the former reaction.

Depending on the inhibitor reactivity, two cases should be considered:

1. Very high inhibitor reactivity. Initiation is almost immediately followed by chain termination and no chain propagation can develop [59]. The autooxidation is delayed until all inhibitor is consumed. Hydroperoxides are depleted in initiation reaction (7) and restored in reaction (62), and it can be assumed that the initial hydroperoxide concentration $[ROOH]_0$ remains unchanged as long as the inhibitor is present. The steady state concentration of peroxy radicals establishes very quickly:

$$\frac{d[RO_2^{\cdot}]}{dt} = k_1[ROOH]_0^2 - k_7[RO_2^{\cdot}][InH] \approx 0 \tag{64}$$

or

$$[RO_2^{\cdot}] = \frac{k_1[ROOH]_0^2}{k_7[InH]} \tag{65}$$

The rate of inhibitor consumption is

$$-\frac{d[InH]}{dt} = k_7[RO_2^{\cdot}][InH] \tag{66}$$

Substituting (65) into (66), we get

$$-\frac{d[InH]}{dt} = k_1[ROOH]_0^2 \tag{67}$$

Upon integration at the initial condition $t = 0$, $[InH] = [InH]_0$:

$$[InH] = [InH]_0 - k_1[ROOH]_0^2 t \tag{68}$$

which means that the consumption of the inhibitor is linear in time and its concentration will be zero at time $t_1 = [InH]_0/k_1[ROOH]_0^2$

The autooxidation is delayed by a period of time which is directly proportional to the initial concentration of the inhibitor and inversely proportional to the rate of initiation. After time t_1, oxidation resumes the autooxidation pace and can be evaluated by Equation (23).

2. Moderate inhibitor reactivity. In this case the inhibitor does not postpone the autooxidation but rather slows it down. The steady state approximation for inhibited autooxidation (reactions (7), (2), (3) and (62)) leads to the following equation for the rate of oxidation:

$$\frac{dX}{dt} = \frac{k_1 k_3}{k_7} \cdot \frac{([A]_0 - X)([B]_0 + X)^2}{[InH]_0} \tag{69}$$

where $[A]_0$, $[B]_0$, and X have the same significance as previously. Integration of Equation (69) and utilization of Equations (21) and (22) gives:

$$\ln \frac{I_t}{I_{max} - I_t} + \left(\frac{I_{max}}{I_t - CX} - \frac{I_{max}}{I_t} \right) = \ln \frac{[B]_0}{[A]_0} + \frac{k_1 k_3}{k_7}$$

$$\cdot \frac{([A]_0 + [B]_0)^2}{[InH]_0} t \tag{70}$$

At low conversion the absolute value of the first term in the left hand part of Equation (70) is much greater than the term in parentheses and Equation (70) can be simplified:

$$\ln \frac{I_t}{I_{max} - I_t} \simeq \ln \frac{[B]_0}{[A]_0} + \frac{k_1 k_3}{k_7} \frac{[A]_0^2}{[InH]_0} t \tag{71}$$

Equation (71) is similar to Equation (23) and the retardation factor (the ratio of the rates of inhibited and uninhibited oxidation) is

$$\sum = \frac{(k_1 k_6)^{1/2}}{k_7} \cdot \frac{[A]_0}{[InH]_0} \tag{72}$$

Thus there seems to be a distinctive difference between highly active and moderate inhibitors. The former are expected to prolong the induction time leaving oxidation rate unchanged, while the latter diminish the rate of oxidation without changing its induction time.

REFERENCES

1. F. L. Gugumus, Polym. Sci. Technol., 26: 17 (1984).

2. G. Kletecka, SPE Trans., 30: 157 (1984).

3. J. B. Howard and H. M. Gilroy, Polym. Eng. Sci., 15: 268 (1975).

4. N. S. Allen, Analysis of Polymer Systems (L. S. Bark and N. S. Allen, eds.), Applied Science, London (1982).

5. G. Lundeen and R. Livingston, Photochem. Photobiol., 4: 1085 (1965).

6. R. F. Vassil'ev, Nature, 196: 668 (1962).

7. R. F. Vassil'ev and A. A. Vichutinskii, Proc. Acad. Sci. USSR Phys. Chem., 145: 588 (1962).

8. V. A. Belyakov and R. F. Vassil'ev, Photochem. Photobiol., 11: 179 (1970).

9. G. E. Ashby, J. Polym. Sci., 50: 99 (1961).

10. M. P. Schard and C. A. Russell, J. Appl. Polym. Sci., 8: 997 (1964).

11. M. P. Schard, Polym. Eng. Sci., 5: 246 (1965).

12. G. A. George, G. T. Egglestone, and S. Z. Riddell, Polym. Eng. Sci., 23: 412 (1983).

13. G. A. George, Polym. Deg. Stabil., 1: 217 (1979).

14. G. A. George, Pure Appl. Chem., 57: 945 (1985).

15. G. A. George, Developments in Polymer Degradation, Vol. 3, Applied Science, London, Chpt. 6 (1981).

16. G. A. George and S. Z. Riddell, J. Macromol. Sci., A14: 161 (1980).

17. G. D. Mendenhall, Angew. Chem. Int. Ed. Eng., 16: 225 (1977).

18. R. A. Nathan, G. D. Mendenhall, M. A. Birts, C. A. Ogle, and M. A. Golub, ACS Adv. Chem. Ser., 169: 19 (1978).

19. G. D. Mendenhall, J. A. Hassell, and R. A. Nathan, J. Polym. Sci., Polym. Chem. Ed., 15: 99 (1977).

20. K. R. Flaherty, W. M. Lee, and G. D. Mendenhall, J. Polym. Sci., Polym. Lett. Ed., 22: 665 (1984).

21. L. Matisova-Rychla, P. Ambrovic, N. Kulickova, J. Rychly, and J. Holcik, J. Polym. Sci. Symp., 57: 181 (1976).

22. L. Matisova-Rychla, Zs. Fodor, and J. Rychly, Polym. Deg. Stabil., 3: 371 (1981).

23. L. Matisova-Rychla, J. Rychly, and M. Vavrekova, Eur. Polym. J., 14: 1033 (1978).

24. J. Rychly, L. Matisova-Rychla, and M. Lazar, J. Polym. Sci., Symp., 57: 139 (1976).

25. J. L. Bolland and G. Gee, Trans. Faraday Soc., 42: 236 (1946).

26. N. Semenoff, J. Chem. Phys., 7: 683 (1939).

27. L. Audouin-Jirackova and J. Verdu, J. Polym. Sci., Polym. Chem. Ed., 25: 1205 (1987).

28. D. F. Marino and J. D. Ingle, Anal. Chem., 53: 1175 (1981).

29. S. Stieg and T. A. Nieman, Anal. Chem., 50: 401 (1978).

30. R. G. Bauman and S. H. Maron, J. Polym. Sci., 22: 1 (1956).

31. S. W. S. McKeever, Thermoluminescence of Solids, Cambridge University Press, Cambridge (1985).

32. M. P. Schard and C. A. Russell, J. Appl. Polym. Sci., 8: 985 (1964).

33. F. Yoshii, T. Sasaki, K. Makuuchi, and N. Tamura, J. Appl. Polym. Sci., 31: 1343 (1986).

34. C. R. Boss and J. C. W. Chien, J. Polym Sci., A-1, 4: 1543 (1966).

35. B. L. Beck, A. A. Hiltz, and J. R. Knox, SPE Trans., 3: 279 (1963).

36. T. Wada, T. Hotta, and R. Susuki, J. Polym. Sci., C-23: 583 (1968).

37. L. Zlatkevich, J. Polym. Sci., Polym. Phys. Ed., 23: 2633 (1985).

38. N. Ya. Rapoport and V. B. Miller, Polym. Sci. USSR, 19: 1758 (1977).

39. T. Mill and G. Montorsi, Int. J. Chem. Kinet., 5: 119 (1973).

40. S. R. Dharwadkar, A. B. Phadnis, M. S. Chandrasekaraiah, and M. D. Karkhanavala, J. Thermal Anal., 18: 185 (1980).

41. E. M. Y. Quinga and G. D. Mendenhall, J. Am Chem. Soc., 105: 6520 (1983).

42. L. Zlatkevich, Polym. Eng. Sci., 24: 1421 (1984).

43. N. Uri, Autooxidation and Antioxidants (W. O. Lundberg, ed.), Interscience, New York (1961).

44. L. Zlatkevich, J. Polym. Sci., Polym. Phys. Ed., 23: 1691 (1985).

45. D. J. P. Harrison and J. F. Johnson, Polym. Eng. Sci., 22: 865 (1982).

46. D. M. Chang, Polym. Eng. Sci., 22: 376 (1982).

47. D. H. Kaelble, The Physical Chemistry of Adhesion, Wiley-Interscience, New York (1971).

48. N. C. Billingham and E. S. O'Keefe, "Oxyluminescence Studies of Polyamides and Polydienes," Proc. IUPAC Polymer 85, Melbourne, Australia, pp. 533—535 (1985).

49. Y. Hori, S. Shimada, and H. Kashiwabara, Polymer, 18: 25 (1977).

50. L. Zlatkevich, Polymer Stabilization and Degradation (P. P. Klemchuk, ed.), American Chemical Society, Washington, D.C. (1985).

51. N. S. Allen and M. J. Harrison, Eur. Polym. J., 21: 517 (1985).

52. J. C. W. Chien and H. Jabloner, J. Polym. Sci., A-1, 6: 393 (1968).

53. V. Pudov and M. Neiman, Neftekhimiya, 3: 750 (1963).

54. A. Tolks and V. Pudov, Izv. Akad. Nauk Lat. SSR, 350 (1973).

55. L. Zlatkevich, Polym. Degrad. Stabil., 19: 51 (1987).

56. L. Zlatkevich, Lubrication Eng., 44: 544 (1988).

57. J. C. W. Chien, J. Polym. Sci., A-1, 6: 375 (1968).

58. W. R. Jones, M. A. Meador, and W. Morales, ASLE Trans., 30: 211 (1987).

59. F. Tudos, Z. Fodor, and M. Iring, Angew. Makromol. Chem., 158/159: 15 (1988).

60. F. Cracco, A. J. Arvia, and M. Dole, J. Chem. Phys., 37: 2449 (1962).

61. A. A. Vichutinskii, Proc. Acad. Sci. USSR Phys. Chem., 157: 663 (1964).

62. R. F. Vasil'ev, Macromol. Chem., 126: 231 (1969).

5

Luminescence Studies of the Photooxidation of Polymers.

NORMAN S. ALLEN   Manchester Polytechnic, Manchester, England

ERYL D. OWEN   University College, Cardiff, South Wales

‹ab. p. 232

I.   INTRODUCTION

The use of luminescence emission and excitation spectra, (fluorescence and phosphorescence) as a sensitive and selective technique for the investigation of various aspects of polymer structure and reactivity has extended over the past 40 years.   It has received an impetus in more recent times as the number and type of polymer materials and composites in commercial use has increased dramatically and an under-standing of the consequent thermal and photochemical degradative and stabilization processes has become more urgent.   While the former has posed problems as far as the processing and fabrication of the polymer at elevated temperatures (>250°C) is concerned, the latter, which is the subject of this account, has imposed severe limitations on the extent of the otherwise potentially large outdoor application.

As these mechanistic investigations have developed, it has become increasingly clear that the substances responsible for the absorption of light and subsequent initiation of the photodegradation processes are often present in very small amounts and their accurate quantita-tive evaluation a matter of some difficulty.   For this application, therefore, the high sensitivity and selectivity of luminescence mea-surements has made the technique a valuable adjunct to the comple-mentary electronic absorption, infrared (vibration-rotation), electron paramagnetic resonance, nuclear magnetic resonance, and electron spectroscopies.

Such is the number, application, and chemical diversity of the polymers currently in commercial use that only a summary of the more prominent and representative examples will be discussed here; emphasis will be placed on behavioral trends which have emerged and which may have more general application as additional data become available.

Somersall and Guillet [1] classified the luminescence emissions of polymers into two main types depending on whether they arose from impurity or "defect" chromophores which are not part of the expected, formal molecular structure (type A), or from the essential, intrinsic structure of the pure polymer (type B). Since much of the disagreement and confusion which has arisen in recent years has been due to an insufficient appreciation of the extent or the effect of such inadventitious additives, no attempt will be made to apply such a classification here. Examples will be selected from a range of polymer types to illustrate the scope and give some indication of the limitations of the technique rather than to make definitive statements about particular mechanistic pathways. In some of the polymers described, the chromophores identified from luminescence studies, whether as initiators or products of the photooxidation process, play a prominent role whereas in others they are less important. In all cases, however, they serve to illustrate the diverse application of the technique.

II. POLYOLEFINS

A. Polyethylene and Polypropylene

Although the ideal structures of polyethylene (PE) and polypropylene (PP) are such that their absorbance of the portion of the terrestial solar spectrum with wavelength greater than 290 nm is so small as to be measurable only with some difficulty, their sensitivity to sunlight was recognized at an early stage in their development. Moreover, this factor still represents some commercial disadvantage in relation to their outdoor application. The photosensitivity is the result of a photooxidation process which introduces various oxygenated chromophores into the polymer backbone as well as promoting a restructuring due to formation of chain scission and crosslinks and which results in catastrophic changes in mechanical properties appearing most obviously as embrittlement. These effects can be minimized, with considerable economic disadvantage, by the addition of photostabilizer combinations which operate by a variety of mechanisms the details of which are still far from beging completely understood. If the precise nature of the light-absorbing species and therefore some detail of the initiating reaction was known with more certainty, it might be possible to reduce its concentration and hence minimize its

effect by suitably treating the polymer. Alternatively, the choice of
an appropriate stabilizing system could be made with more confidence
in the light of such data. The very low levels of the possible initiating chromophores, however, make their accurate quantification and
assessment of their relative importance difficult.

Several possible active chromophores have been suggested at
various times, their relative importance depending on the factors
which determine the amount of radiation absorbed, i.e., their concentration and molar extinction coefficient (ε_λ), and the quantum
yield of the photooxidation process. Those now believed to be the
most important, not necessarily in order of efficiency, are discussed
below.

1. Peroxide (R—O—O—R) and Hydroperoxide (R—O—O—H) Groups

Peroxide and hydroperoxide groups have long been implicated in the
photooxidation of PE and PP [2,3], the initiation beginning with
scission of the —O—O— bond with the formation of R—O and OH free
radicals. The low values of $\varepsilon(\lambda)$ for these chromophores means that
although their importance as primary absorbers may be less than
that of other chromophores in the early stages, the high quantum
yield of the photocleavage process indicates that they may be acti-
vated by a sensitized process involving energy transfer from some
other absorbing species. The importance of energy transfer processes in which radiation aborbed by one chromophore is then transferred
efficiently to another, more labile species has only become apparent
in relatively recent times. Since energy transfer is a primary photochemical process and competes with other radiative deactivation modes,
luminescence quenching is a convenient technique for assess ing its
importance. An example will be described in more detail in Section
II.C., which is concerned with polystyrene photooxidation. At this
stage it is sufficient to indicate that energy transfer may play some
part in explaining the different relative importance of —OOH groups
reported for PP and PE photolysis [4,5].

2. Carbonyl Groups ($>$C=O) and ($\sim$(CH=CH)$_n$—C=O)

Carbonyl groups introduced into polyolefins during the polymerization
and/or processing stage have often been implicated as photoinitiators
of the oxidation process [6]. The mechanism of ketone photolysis
in low molecular weight ketone model compounds has been exhaustively
investigated and the results obtained shown to be largely applicable
to macromolecular systems. Both Norrish type I and type II processes result in chain scission, and in addition the type I process
[reaction (1a)] results in the formation of free radicals which initiate
the oxidative reaction.

$$\sim CH_2CH_2CH_2\underset{\underset{O}{\parallel}}{C}CH_2\sim \xrightarrow{\text{Type I}} \sim CH_2CH_2 \cdot + \underset{\underset{O}{\parallel}}{C} + \cdot CH_2\sim$$

Type II $\downarrow$

$$\underset{\underset{\underset{H}{}}{\overset{\displaystyle \sim C}{\underset{H}{\diagdown}}}}{\overset{\displaystyle CH_2CH_2}{\diagup \diagdown}} \underset{\underset{H}{}}{\overset{\displaystyle CH_2\sim}{O \diagup}} \qquad \sim CH=CH_2 + \underset{\underset{O}{\parallel}}{\overset{\displaystyle H_3C}{\diagdown}} C\cdot CH_2\sim \qquad (1)$$

Luminescence has been used extensively to monitor the concentration and reactions of carbonyl groups since the phosphorescence spectra of these molecules makes their detection relatively easy. Results which are typical of those obtained from luminescence measurements have been obtained from a study of the correlation of the luminescence of thermally oxidized PP with its light stability [7]. The phosphorescence emission spectrum of PP [λ(ex) = 280 and 330 nm], with maxima at 415, 455, and 485 nm and the excitation, [λ(em) = 460 nm], with maxima at 260, 280, and 330 nm are shown in Figure 1.

Heating PP powder in air at 130°C for various lengths of time up to 45 min caused a reduction in the intensity of both fluorescence and phosphorescence spectra. When heating was continued for longer periods, however, the phosphorescence emission at wavelengths longer then 450 nm increased and a new excitation maximum appeared at 310 nm. No such dramatic change occurred in the fluorescence spectrum over the same period. Samples which had been heated for longer then 60 min showed a greatly increased sensitivity to light relative to those heated for shorter periods as measured by greatly increased embrittlement and a rapid increase in carbonyl index. Irradiation was carried out in a Xenotest, accelerated weathering cabinet. The species produced had phosphorescence emission maxima at 310 nm which have been attributed to saturated ketonic or aldehyde chromophores.

The relative importance of the role played by unsaturated ketonic structures, principally enones and dienones, in the initiation of the photooxidation of PE and PP has proved difficult to quantify unambiguously and has been a matter of some dispute in the past. Measurements made by Allen et al. [8] showed that the fluorescence excitation spectra of commercial PE and PP as well as poly(4-methylpent-1-ene) showed a close similarity with the absorption spectra of some typical aliphatic model enones (Figure 2) and supported in general terms the earlier work of Charlesby [9]. By analogy with well-documented data [10,11] on the photochemistry of α,β-unsaturated carbonyl compounds, these authors concluded that they played a significant role in the photooxidative degradation of PE and PP.

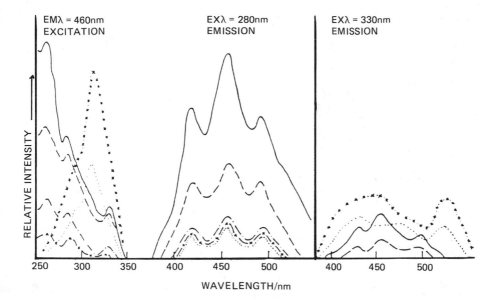

FIGURE 1   Changes in the phosphorescence emission and excitation spectra of polypropylene powder after heating for 0 min (——); 15 min (---); 30 mn (-·-·); 45 min (-x-x); 60 min (····); and 75 min (·x·x) in air at 130°C.   (Reproduced from Ref. 7 with permission from John Wiley and Sons, New York.)

During this process the initiators are converted into saturated ketonic/aldehydic groups which are in turn converted into nonluminescent products such as carboxylic acids.

An alternative suggestion by Carlson and Wiles [12] was that polynuclear hydrocarbons like naphthalene, anthracene, or phenanthrene were initiators of the photooxidation, acting by sensitizing the production of molecular oxygen in its lowest singlet excited state.

$$PNH(S_0) + h\nu(\lambda > 290 \text{ nm}) \longrightarrow PNH(S_1) \rightsquigarrow PNH(T_1) \qquad (2)$$

$$PNH(T_1) + O_2(^3\Sigma_g) \longrightarrow PNH(S_0) + O_2(^1\Delta_g) \qquad (3)$$

The suggestion arose following experiments in which $^1O_2$ produced in a microwave discharge reacted with model unsaturated compounds like squalane to produce hydroperoxides with high efficiency.   No such reactivity was observed toward saturated molecules.   Polypropylene films which had been pretreated with oxygen, however, became sensitized toward photooxidation presumably due to the conversion of allylic unsaturation, occurring at random sites in the polymer chain,

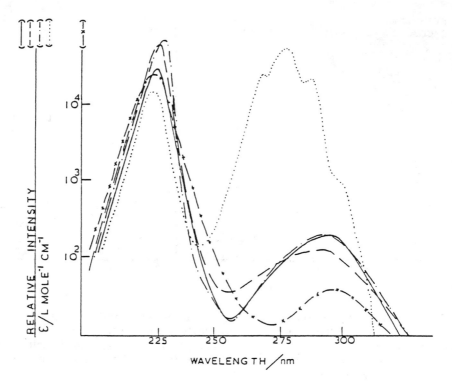

FIGURE 2 Comparison of the fluorescence excitation spectra of polypropylene (——), poly(4-methylpent-1-ene) (----), and low-density
polyethylene (-·-·) films with the absorption spectrum of pent-3-ene-
2-one (-x-x) and the fluorescence excitation spectrum of naphthalene
(···) in $\underline{n}$-hexane (10^{-5} M). (Reproduced from Ref. 8 with permission from John Wiley and Sons, New York.)

into photolabile hydroperoxide groups.

$$\sim CH_2-CH=CH-\overset{|}{\underset{|}{C}}\sim + {}^1O_2 \longrightarrow \sim CH_2-\underset{\underset{O-H}{\overset{|}{O}}}{\overset{|}{CH}}-CH=CH\sim \qquad (4)$$

Ni(II) chelate molecules [13] as well as aromatic diammines which had
previously been shown [14] to have a protective effect on polymers
against the effects of light were efficient $^1O_2$ quenchers. In addition,
the deliberate introduction of PNH molecules into PP markedly increased
the rate of photooxidation. These conclusions received some support

from the work of Osawa and Hiroda [15], who claim that on extraction of the polymers with hexane solvent the fluorescence emission disappeared completely, the phosphorescence became very weak, and the emissions were transferred to the extract. Photolysis reduced the emission intensity of the extract and of the unextracted films.

The considerable role played by α,β-unsaturated carbonyl compounds in the photooxidation of polybutadiene, described in Section II.B, should be taken into consideration before deciding on their relative role in the cases of PE and PP.

3. Other Photosensitizers

Other possible photosensitizing species which have been suggested from time to time include traces of impurity transition metal ions or remaining fragments of the catalyst used in the polymerization. Some evidence has been offered for the involvement of olefin—O_2 charge transfer complexes but they probably are much less important in the case of PE and PP than in other systems. One closely related source of photoinstability which will be mentioned here is that arising from the addition of pigments like TiO_2, which is amenable to investigation using luminescence methods.

The effects of various loadings of TiO_2 on the phosphorescence and subsequent carbonyl formation on the photooxidation of PE and PP was investigated by Allen and McKellar [16]. The presence of the anatase form of the pigment quenched the phosphorescence of PE (470 nm) and PP (460 nm) but, unlike PP, PE showed some residual emission at $\lambda < 470$ nm. As the anatase loading was increased the intensity of the anatase emission at 540 nm became predominant. The rutile form of the pigment also reduced the phosphorescence intensity of both polymers, but while both polymers showed a measurable increase in photosensitivity for anatase loadings greater than 0.5%, this was not the case for the rutile form, which suggests that it may play only a passive (i.e., screening) role. The additional effect of o-hydroxybenzophenone or Ni(II) chelate photostabilizers on the photooxidation process was complex and difficult to interpret with confidence but seemed to indicate that excited state quenching was an important component in some cases.

B. Polybutadiene

Polybutadiene (PBD) has found extensive application as a component of the terpolymer ABS (acrylonitrile-butadiene-styrene) in spite of the high susceptibility to photooxidation which is introduced by the presence of the three unsaturated structural units (I-III) which make up the polymer structure:

$$\sim CH_2 \overset{CH=CH}{\underset{(I)}{\quad}} CH_2 \sim \qquad \sim CH_2 \overset{CH=CH}{\underset{\sim CH_2}{\quad}} \overset{CH_2 \sim}{\underset{(II)}{\quad}} \qquad \sim CH_2 CH \sim \overset{CH}{\underset{(III)}{\quad}} \overset{CH}{CH_2}$$

Early investigations of the thermal and photochemical processes concluded that both led to the formation of similar products and that
the mechanisms may well be the same [17]. Both processes are
autocatalytic, and carbonyl and hydroxyl groups are formed with con-
current loss of unsaturation. Luminescence measurements have made
a useful contribution to understanding the photooxidative process.
Phosphorescence emission and excitation spectra of the polymer after
thermal and photooxidation are shown in Figure 3. The polymer was
initially nonluminescent but after oxidative thermal degradation,
phosphorescence emission was observed with λ(max) at 475 nm, and
a shoulder at 520 nm and corresponding excitation spectrum with
λ(max) = 310 nm. Following photooxidation, similar spectra were
observed with an emission maximum at 520 nm and the corresponding
excitation maximum at 340 nm. This emission had a shoulder at

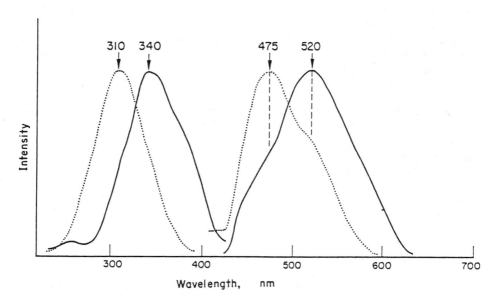

FIGURE 3 Phosphorescence emission and excitation spectra of photo-
(——) and thermally (···) oxidized polybutadiene. (Reproduced from
Eur. Polym. J., 10, 593 (1974), with permission from Pergamon Press,
Ltd., UK.)

475 nm which in some cases was resolved into a separate peak. The
peak at 475 nm has been attributed, using infrared and ultraviolet
analysis, to the presence of one or more saturated aldehydes, the
phosphorescence emission spectrum closely resembling that of model
saturated aldehydes. That at 520 nm is due to one or more α, β-
unsaturated carbonyls. That saturated ketone phosphorescence is
not observed (when it has been shown that such molecules are present
in oxidized films) is a consequence of their small phosphorescence
efficiency relative to that of the corresponding aldehydes. The
phosphorescence of the thermally degraded film owes most of its
intensity to the emission of the one or more α, β-unsaturated carbonyls
produced giving rise to the maximum at 520 nm.

The fact that the phosphorescence of the saturated and unsaturated
aldehydic group(s) present are of approximately equal intensity when
it has been shown that the α, β-unsaturated carbonyls are formed
almost to the exclusion of the saturated variety is again a consequence
of the greater phosphorescence efficiency of the saturated aldehydes.

The realization that α, β-unsaturated carbonyls were the initiators
of the photooxidation process led Beavan and Phillips [10] to investigate the possibility that substances which could quench the excited
states of these molecules with high efficiency might prove to be
effective stabilizers for the photooxidation process. This would be
an attractive alternative to the more traditional approach of incorporating inert substances with high extinction coefficients which pre-
ferentially absorb the incident radiation, thus inhibiting the photooxidation process. If the reasonable assumption is made that it is
the triplet states of the α, β-unsaturated carbonyls which are photochemically active, then those substances with triplet levels lower
than about $2.2 \times 10^{-6} \ m^{-1}$ could act as acceptors in an exothermic
triplet—triplet energy transfer process and thus act as effective
stabilizers for PBD photooxidation, i.e.,

$$^3(\alpha, \beta\text{-unsat} > C{=}O)(T_1 = 2.2 \times 10^{-6} \ m^{-1}) + Q(T_1 < 2.2 \times 10^{-6} \ m^{-1})(So)$$

$$\longrightarrow (\alpha, \beta\text{-unsat} > C{=}O(So) + {}^3Q(T_1) \qquad (5)$$

The triplet energies of 26 commercial ultraviolet stabilizers were
measured from the onset of the phosphorescence spectrum. The two
benzotriazole compounds Tinuvin P (IV) and Tinuvin 327 (V) have low
enough triplet energies to act as acceptors of the energy of the α, β-
unsaturated carbonyl donor and they prove to be the most efficient
additives in commercial use. Moreover, they proved to be efficient
quenchers of the phosphorescence of the model α, β-carbonyl compound
VI by energy transfer, the process giving rise to the sensitized
phosphorescence of the additive (Figure 4).

(IV)

(V)

(VI)

It is apparent that the data concerning the role of α, β-unsaturated carbonyl compounds in the photooxidation of PBD may also be relevant to the photooxidation of PE and PP which has been discussed already. Any such consideration should pay particular attention to their relative concentrations, extinction coefficients, and luminescence quantum yields when attempting to assess their relative importance.

C. Polystyrene

Various aspects of the photooxidation of polystyrene (PS) have been studied continuously for at least 30 years, ever since its susceptibility to photocoloration and embrittlement became apparent. Early papers dealt mainly with structural changes and were studied by infrared and ultraviolet spectroscopy and the analysis of volatile photoproducts. In more recent times luminescence studies have played a large part in establishing the details of the photoinitiation process. An important step toward an understanding of the mechanistic detail was made by Grassie [18], who showed by comparing the photolysis of the polymer in oxygen with that under vacuum that the reaction did not show the autocatalytic behavior which is typical of other polymers. It became clear at an early stage, therefore, that some reliable indication was required of the nature of the chromophores responsible for photoinitiation. George and Hodgeman [19] investigated the role played by "in-chain" peroxides which had been suggested as possible contenders by Lawrence and Weir [20]. They prepared polystyrene at 60°C in the presence of various partial pressures of oxygen (10^{-4} to 10^{-1} mm Hg) and showed by quantitative phosphorescence analysis that they contained lower levels of phenylalkylketone groups than is normally the case for commercial PS samples. On photolysis at 77K followed by warming to 290K there was a rapid increase in the phosphorescence emission at 398 nm (Figure 5),

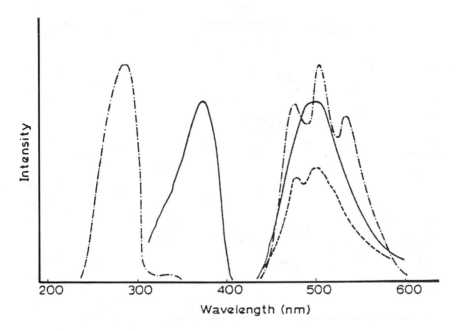

FIGURE 4   (left) Phosphorescence emission and excitation spectra of
Tinuvin P ($-\cdot-\cdot$) and VI (—). (right) Phosphorescence emission
of a mixture of Tinuvin P and VI (- - -) [λ(ex) = 380 nm]. (Reproduced from Ref. 10 with permission of Elsevier Sequoia, UK.)

attributed to phenylalkylketone end groups, but quite different
behavior was observed for samples which had been hydroperoxidized.
They concluded that in-chain peroxides may be regarded as precursors
of the initiators, phenylalkylketones, in PS prepared or processed at
high temperatures and may even survive and act as direct photoinitiators in samples prepared at lower temperatures. The results of

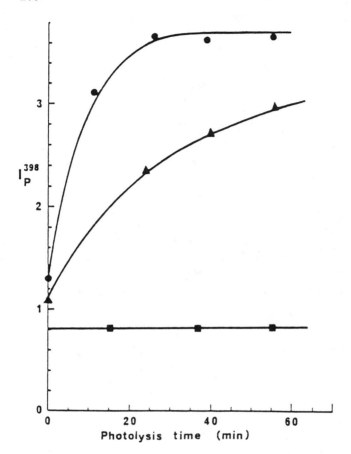

FIGURE 5   Changes in phosphorescence emission intensity at 398 nm
on photolysis for PS pretreated as follows: hydroperoxidized (■);
photolyzed at 77K under 10^{-1} mm Hg of oxygen warmed to 290°K
then recooled to 77°K (●); as (●) except for 10^{-3} mm Hg of oxygen
(▲).   (Reproduced from Ref. 19 with permission from John Wiley and
Sons, New York.)

Geuskins [21], however, led to a different emphasis.   Using light of
253.7 nm as the source in the presence of 600 torr of oxygen, it was
shown that the rate of oxygen absorbance remained constant with
time and occurred with a quantum yield, (ϕ_{-O_2}) of 2.7×10^{-2}, which
was larger than that for hydroperoxide formation ($\phi_{ROOH} = 8.6 \times$
10^{-4})   measured by infrared absorbance in the range 3200 to 3500
cm^{-1}.   This difference was attributed to the photolysis of hydroperoxides to form secondary photoproducts which include carbonyl

groups of type VII, formed from initiation of oxidation at the secondary carbon on the polymer backbone, as well as acetophenone groups

(VII)

produced as a consequence of main-chain scission.   George and Hodgemen [22] showed, using careful quantitative phosphorescence spectroscopy and viscosity measurements, that a typical commercial polymer sample contains 0.14 carbonyl group per number average chain and that the number of acetophenone groups and the number of chain scissions are quantitatively related.   A likely mechanism involves the decomposition of a tertiary hydroperoxide group with elimination of water.

(7)

However, comparison of the increases in absorbance which occurred at 240 nm as photooxidation proceeded with the corresponding changes in the infrared region indicated that acetophenone end groups were only partly responsible for the change and must be accompanied by a contribution from other chromophores.   The most likely on the basis of a qualitative examination of the ultraviolet spectra are the unsaturated chain ends produced in reaction (7). A quantitative correlation of their absorbance and concentration is made difficult by their reaction with oxygen as well as their conversion into longer polyenes, which are responsible for the progressive extension of the tail of the absorption into the visible region.   Comparison with model compounds [23] suggests that absorption maxima should be observed at 280 nm for n = 1 (diene), 310 nm for n = 2 (triene), and 340 nm for n = 3 (tetraene), and absorptions are observed in these regions.   Formation of polyenes occurs following energy transfer from phenyl groups in PS monomer (S) in a way which is analogous to the PS → monomer (S) system [24].   The excited end groups then decompose with elimination of hydrogen.

$$\sim(CH_2\ CH)_n\sim\ +\ h\nu\ \rightarrow\ \sim(CH_2\ CH)_{n-1}\ CH_2\ CH\sim \qquad (8)$$

$$-CH\sim\ +\ CH_2\overset{-}{C}\ CH_2\ CH\sim\ \rightarrow\ (CH_2\overset{-}{C}\ CH_2\ CH)^*\sim$$

$$\qquad\qquad\qquad\qquad\qquad\qquad\qquad\qquad\qquad (9)$$

$$CH_2\overset{-}{C}\ CH=C\sim\ +\ H_2$$

Strong support for the ultraviolet data indicating the formation of polyenes was obtained [21] from changes in the fluorescence spectra which accompanied photooxidation (Figure 6). The excimer fluorescence (λ_{max} = 330 nm), which was observed initially, decreased in intensity as photooxidation proceeded and was replaced by a broad emission extending to about 550 nm with maxima around 350, 400, 450, and 500 nm. The closest model systems for which comparable fluorescence data are available is the series of α,ω-diphenyl polyenes (VIII) [25]. Fluorescence emission maxima of these molecules shift progressively to longer wavelength as n increases in a way which is closely analogous to that observed in the PS photoproduct

$$\langle\bigcirc\rangle(CH=CH)_n\langle\bigcirc\rangle$$
$$(VIII)$$

In addition to their utility in the identification of photoinitiating sites in PS photooxidation, luminescence techniques have also proved invaluable in the investigation of energy transfer between chromophores during the early stages of the process. The intensity of the excimer fluorescence of PS decreases rapidly during the early stages of photooxidation (Figure 7) because energy is transferred from the excited phenyl group to oxidation products as they are formed instead of migrating to excimer sites. This transfer becomes increasingly efficient until finally excimer fluorescence is undetectable. It sometimes happens when energy transfer occurs in this way that the emission of the acceptor becomes visible as that of the donor decreases. During the photooxidation of PS, however, the only new emissions which are detectable are the polyene fluorescence and the acetophenone end group phosphorescence, although the former make little

i

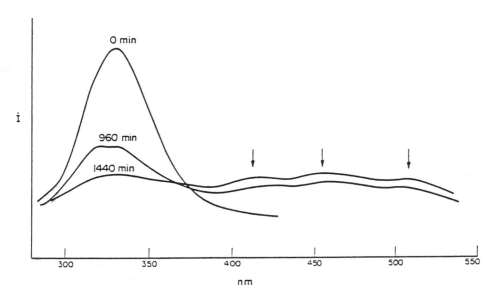

FIGURE 6 Fluorescence emission of PS film [λ(ex) = 254 nm], irradiated (λ = 253.7 nm) in oxygen. (Reproduced from Ref. 21 with permission from Pergamon Press, Ltd., UK.)

or no contribution to the initiation of the photooxidation. Further, although it is clear from ultraviolet and infrared data that the concentration of acetophenone end groups increases regularly as photooxidation proceeds, the emission arising from these chromophores increases at first but then falls to a lower level (Figure 8). The most likely explanation of this phenomenon is that as the photooxidation proceeds, energy is transferred progressively from the triplet excited acetophenone to some other nonluminescent chromophore whose concentration increases with time in such a way that the transfer becomes increasingly efficient. The results indicate strongly that hydroperoxides may be the species in question and are nonluminescent because of the high quantum yield of O—O bond scission. Studies using films of PS containing cumene hydroperoxide as an energy acceptor support these ideas by showing conclusively that this additive efficiently quenches the excimer luminescence of PS while sensitizing the photooxidation reaction [26].

The considerable part played by luminescence techniques in clarifying the details of the PS photooxidation process may be summarized by reference to the fluorescence of PS excimer and polyene end group chromophores and the phosphorescence of phenylalkylketone groups. Their role in energy transfer processes that lead to the

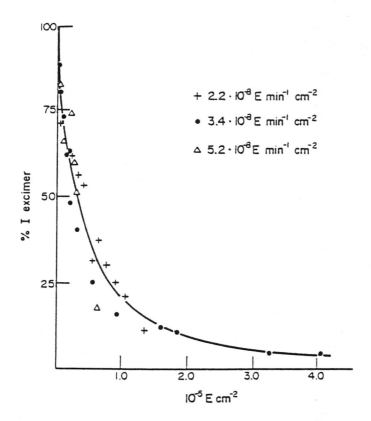

FIGURE 7 Decay of eximer fluorescence [λ(em) = 330 nm] for PS
films irradiated (λ = 253.7 nm) under oxygen. (Reproduced from
Ref. 26 with permission from Pergamon Press, Ltd., UK.)

sensitized decomposition of the hydroperoxides that are the primary
initiators of the process is now well established in both real and
model systems.

III. POLYVINYLCHLORIDE

The primary product which arises from the degradation of polyvinylchloride (PVC), carried out in the absence of oxygen, whether
thermally, photochemically, or by ionizing radiation, is a distribution
of conjugated polyene sequences of various lengths produced by a
dehydrochlorination process which may be written:

$$\sim(CH_2CHCl)\tilde{_n} \longrightarrow \sim(CH=CH)\tilde{_n} + nHCl \tag{10}$$

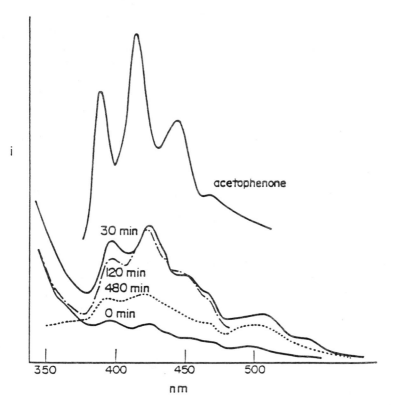

i

FIGURE 8   Phosphorescence emission from PS films irradiated (λ = 253.7 nm) under oxygen.   (Reproduced from Ref. 26 with permission from Pergamon Press, Ltd., UK.)

Immediate consequences of this reaction become apparent when values of n increase to the point where absorptions of the polyenes extend into the visible region of the spectrum and the polymer becomes colored.   For this reason, measurements of absorption in the visible and ultraviolet regions of the spectrum have been widely used as an indication of the extent of degradation.   In the presence of molecular oxygen, photooxidation of the polymer occurs which leads to chain scission and crosslinking, and results in a deterioration in the polymer's mechanical and electrical properties [27].

Details of initiation and propagation of the photooxidation process have been discussed at length [28], and it is generally agreed that the presence of various structural features in the polymer, which occur to different extents depending on the polymerization conditions, is a critical factor.   Recent improvements in analytical techniques have made it possible to identify such structures and measure their

low concentrations with greater certainty. Several possible candidates
have been suggested from time to time but it now appears that the
three most effective structures are (1) carbonyl groups, (2) unsaturated polyene structures, and (3) peroxides, although others may
play a significant role under some conditions. Different workers
have come to different conclusions regarding the relative importance
of the various possible photosensitizers [28].

When PVC films were irradiated under photooxidative conditions,
the degradation was confined to a thin surface layer which became
highly colored and intensely fluorescent [29]. Degassed THF solutions of the photodegraded PVC showed a broad structured fluorescence emission [λ(ext) = 320 nm] with clearly defined features at
360, 380, 405, 429, and 455 nm. Excitation spectra corresponding
to emissions at 500 and 380 nm clearly show the presence of at least
two emitting species, the major one with absorption maxima at 372
and 355 nm and the second with less well-defined maxima at 335 and
300 nm (Figure 9) [30]. The most likely explanation is that the
emission originates from a mixture of at least two polyene sequences
which, if of the type $CH_3(CH{=}CH)_nCH_3$, would correspond to n =
6 to 8 by assimilation of the data of Sondheimer et al. [31], Palma
and Carenza [32], Bengough and Varma [33], and Shindo and Hirai
[34]. The emissions were closely similar, both in the positions of
the maxima and in shape, to those of THF solutions of PVC degraded
thermally in the presence of oxygen (Figure 10), and also to those
of chemically degraded PVC prepared by the method of Shindo et al.
[35].

The emissions from degraded PVC were first attributed to the
distribution of polyene sequences described in Equation 10. Interest
in these polyenes arises not only from the fact that they are a prime
source of the undesirable coloration of degraded PVC but are also
sensitizers of the degradation process. Balandier and Decker [36]
irradiated (250 to 300 nm) THF or DCE solutions of PVC as well as
PVC films, and measured quantum yields of dehydrochloration, ϕ(HCl),
main-chain scission, ϕ(MCS), and crosslinking, ϕ(CL), under N_2 and
O_2 atmospheres. Data obtained showed that the yields of all processes
were considerably increased by the presence of O_2 and that the auto-
accelerating rate which was observed resulted from increased light
absorption by the polyenes.

The suggestion that the emissions could not be accounted for by
the presence of polyene sequences alone but contained contributions
from closely related oxygenated chromophores emerged from several
sources. The careful and extensive investigations of Becker [37—39],
for example, documented the positions and relative intensities of many
polyenes of retinol, -al and -one type. Data extracted from this
work which illustrate the trends in λ(Emn) and λ(abs) as the polyene
length (n) changes are given in Figure 11. Values are included for
PVC which immediately indicate that they cannot be fitted satisfactorily

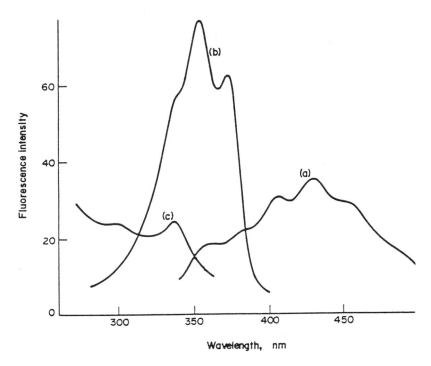

Wavelength, nm

FIGURE 9 Fluorescence emission (a) [λ(ex) = 320 nm] and excitation
spectra (b) and (c) [λ(em) = 450 and 380 nm] of degassed THF
solution (0.01%) of photodegraded PVC. (Reproduced from Ref. 30
with permission from Pergamon Press, Ltd., UK.)

into either linear relationship and therefore cannot be identified
firmly with any of these structures.

The assignement by Maruyama [40] using emission and excitation
spectra of conjugated polyene systems produced by heating polyvinyl
alcohol (PVA) film is also relevant to the situation in PVC. When
freshly prepared PVA film with absorption maxima at 280 and 330
nm, which had been assigned to the $\sim(CH{=}CH)_2CO\sim$ and $\sim(CH{=}
CH)_3CO\sim$ chromophores, respectively, were heated, the absorptions
increased in intensity and broad emission developed around 500 nm
[λ(ex) = 325 or 358 nm] and 590 nm [λ(ex) = 480 nm]. The corresponding excitation spectra were highly structured and were located
in the 300- to 400-nm region corresponding to the tetraene and
pentaene structures assigned previously [41] (Figure 12). The
broad emission at 590 nm excited by 480-nm light was most intense
in spite of the fact that the absorption intensity at 480 nm was
weaker than that at the other excitation wavelengths.

THE INSTITUTE OF METALS

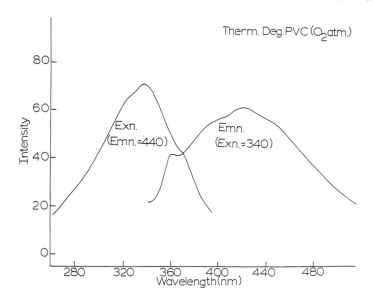

FIGURE 10 Fluorescence emission [λ(ex) = 340 nm] and excitation [λ(em) = 440 nm] spectra of a THF solution of thermally degraded PVC.

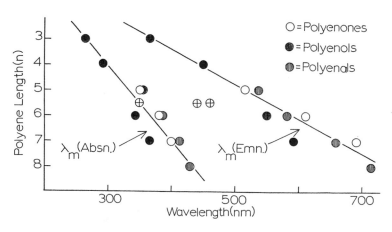

FIGURE 11 Variation of absorption and fluorescence emission maxima with polyene length (n) for a series of polyenones, polyenols, and polyenals. Values for PVC are indicated by ⊕.

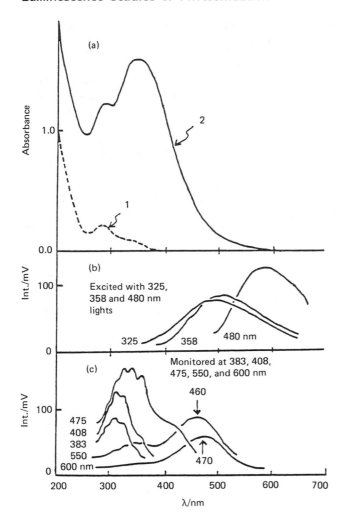

FIGURE 12 (a) Absorption, (b) emission, and (c) excitation spectra of heat-treated PVA film (a) 1 before heating, (a) 2 after heating. (Reproduced from Ref. 40 with permission from the Chemical Society of Japan.)

When the heat-treated film corresponding to Figure 12a(2) was digested in an aqueous solution of NaOH—NaBH$_4$, the absorption maxima at 280 and 340 nm were replaced with the fine-structured absorptions shown in Figure 13a, indicating that the diene and triene carbonyls had been reduced to diene and triene by reduction of the carbonyl group. In addition, neither emission at 590 nm nor excitation at 460 nm could be observed. Similar results, differing only in small detail, were obtained for films which had been digested in CH$_3$COONa solution (Figure 14) or HCl solution (Figure 15), in addition to the procedures used for the film of Figure 12.

The conclusions which can be drawn from these results for PVA and, by close analogy, be applied with confidence to PVC are that the absorptions and emissions observed arise from a mixture of polyene ($\sim(CH=CH)_{\overline{n}}$) and polyene carbonyl ($\sim(CH=CH)_n\sim$) structures, detailed descriptions of which are given in Tables 1 and 2.

TABLE 1 Absorption Maxima attributed to Polyene ($\sim(CH=CH)_n\sim$) Structures in Heated PVA Films

n	λ(nm)			n	λ(nm)
2	242	234	226	9	442
3	285	273	262	10	465
4	325	310	295	11	486
5	358	342		12	503
6	366			13	523
7	390			14	540
8	418				

TABLE 2 Absorbtion and Excitation Maxima Attributed to Polyene Carbonyl ($\sim(CH=CH)_nCO\sim$) Structures in Heated PVA Films

n	λ(nm)	n	λ(nm)
2	280	5	420(s)
3	340, 350(s)	6	450, 460
4	380(s)	7	470

Note: s = shoulder.

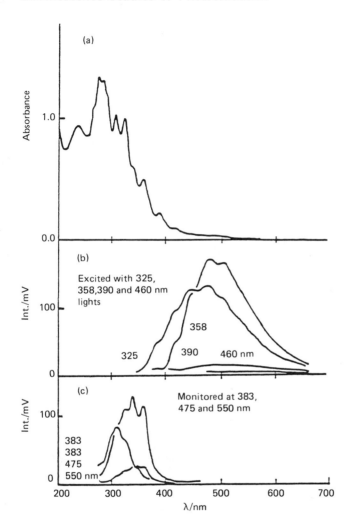

FIGURE 13   (a) Absorption, (b) emission, and (c) excitation spectra of film of Figure 12 after digestion in 0.1 M NaOH-NaBH₄ solution. (Reproduced from Ref. 40 with permission from the Chemical Society of Japan.)

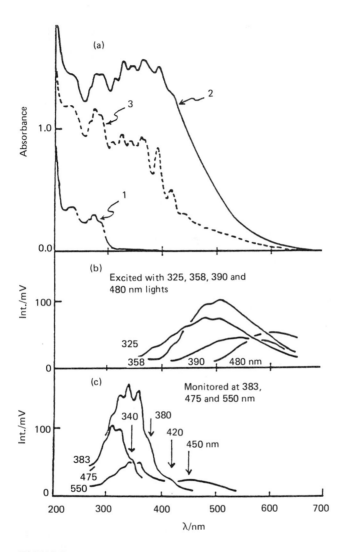

FIGURE 14 (a) Absorption, (b) emission, and (c) excitation spectra
of film of Figure 12 with additional digestion in CH_3COONa solution.
(a) 1 before heating, (a) 2 after heating, (a) 3 after digestion of (a)
2 in NaOH=NaBH$_4$. (Reproduced from Ref. 40 with permission from
the Chemical Society of Japan.)

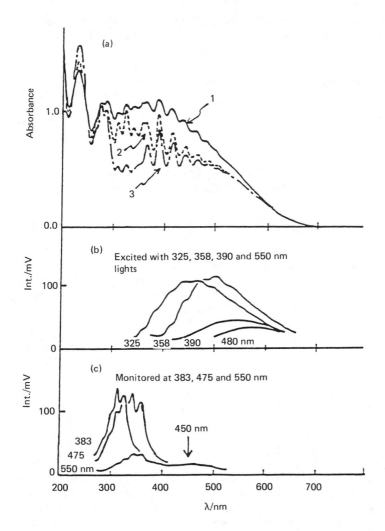

FIGURE 15   (a) Absorption, (b) emission, and (c) excitation spectra of film of Figure 12 with additional digestion in HCl solution, (a) 1 after heating, (a) 2 and (a) 3 after digestion for 1 day and further 3 days in NaOH-NaBH$_4$ solution.   (Reproduced from Ref. 40 with permission from the Chemical Society of Japan.)

Mechanisms by which carbonyl chromophores are photochemically and/or thermally introduced into the structure of PVC have been reviewed in detail by several authors [27,28,42]. These will not be repeated here except to say that they involve initiation by the radicals A, B, and C below formed by abstraction of different hydrogen atoms from the polymer. These are converted by reaction with

$$\sim CH_2\dot{C}HCH_2 \sim \qquad \sim CH_2\dot{C}ClCH_2 \qquad \sim CHCl\dot{C}HCHCl\sim$$

$$\text{(A)} \qquad\qquad\qquad \text{(B)} \qquad\qquad\qquad \text{(C)}$$

oxygen into secondary peroxy radicals $AO_2^{\cdot}$ and $CO_2^{\cdot}$ and the tertiary peroxy radical $BO_2^{\cdot}$, which play decisive roles in the subsequent propagation processes.

IV. POLYAMIDES

Much of the information relating to the photooxidation of polyamides has been derived from the use of luminescence techniques to study the thermal and photochemical oxidation of the model amides IX and X, neither of which shows any significant luminescence before heating or after heating under vacuum [43,44]. However, when these

$$CH_3(CH_2)_3NHCO(CH_2)_4CONH(CH_2)_3CH_3; \quad \underline{N},\underline{N}'\text{-}\underline{di}\text{-}\underline{n}\text{-butyladipamide}$$

$$\text{(IX)}$$

$$CH_3(CH_2)_4CONH(CH_2)_6NHCO(CH_2)_4CH_3; \quad \underline{N},\underline{N}'\text{-dicaproylhexamethy-}$$
$$\text{lenediamine}$$

$$\text{(X)}$$

compounds were subjected to thermal oxidation under relatively mild conditions (180°C), they formed products whose fluorescence emission and excitation spectra closely resembled those of the polyamide nylon 6,6 (Figures 16 and 17). Analysis of the vibrational structure of the phosphorescence from these oxidized model amides led Allen and McKeller [45] to conclude that the emissions originated from the triplet $n\pi^*$ states ($^3n\pi^*$) of α,β-unsaturated carbonyl compounds rather than from the amide chromophore itself, as had once been suggested [46,47], or from cyclic carbonyl structures such as XI derived from polyamides [48,49]. Support for this suggestion was obtained from lifetime studies of the emissions from a variety of nylon polymer chips (Table 3). Although the lifetimes are longer than those of model cyclic dienones, there is a strong correspondence between the lifetime and the extent of the corresponding hypsochromic and bathochromic shifts of $^3n\pi^*$ and $^3\pi\pi^*$ states, respectively, in the

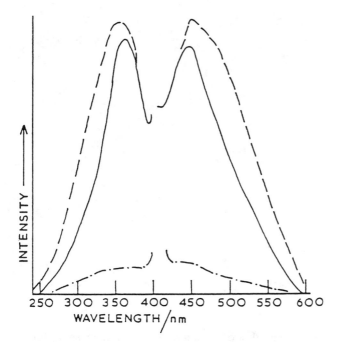

FIGURE 16   Comparison of fluorescence excitation and emission
spectra of nylon 6,6 chip (——), with model amides before (-·-·)
and after (---) heating in air at 180°C.   (Reproduced from Ref. 44
with permission from John Wiley and Sons, New York.)

TABLE 3   Phosphorescence Characteristics of Some Polyamide Polymers

Polyamide	λ(ex)(nm)	λ(em)(nm)	Lifetime (τ)(sec)
Nylon 6 chip	282	390(s),420,455(s)	1.7,1.6,1.1
	300	460	0.80
	310	470	0.58
Nylon 6,6 chip	295	410	2.1
	310	415,460	1.6,0.6
	315	420,460,480	0.7,0.25,0.2
Nylon 12 chip	268,286(s)	363(s),410	1.0
	300	465	0.6
	310	475	0.6

Note: s = shoulder.

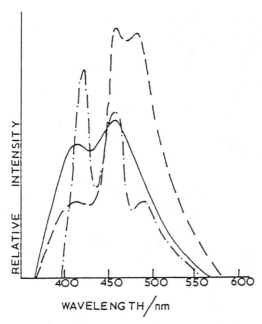

FIGURE 17 Comparison of the phosphorescence emission from nylon
6,6 chip [λ(ex) = 310 nm] (——) (Sx = 300) and [λ(ex) = 315 nm]
(---) (Sx = 450) with that from thermally oxidized model amides
[λ(ex) = 273 nm] (-·-·). (Reproduced from Ref. 45 with permission
from Elsevier Sequoia, UK.)

(XI)

strongly hydrogen-bonded matrixes such as nylon 6,6. These
important structural and environmental factors are difficult to quantify
but lend general support to the suggestion that α,β-carbonyl compounds are the source of the luminescence.

The effects of thermal and photochemical oxidation on the phosphorescence spectra of fresh nylon 6,6 polymer in the chip form are
compared in Figure 18 for two sets of excitation and emission wavelengths. Thermal oxidation at 180°C causes a decrease in the

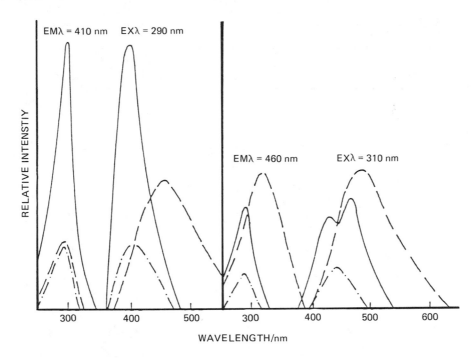

FIGURE 18   Phosphorescence spectra of nylon 6,6 film before (——)
and after thermal (---) and photooxidation (-·-·).   (Reproduced
from Ref. 45 with permission from Elsevier Sequoia, UK.)

intensity of the phosphorescence emission [λ(ex) = 290 nm], with
the appearance of a new, longer wavelength maximum.   This has
been attributed to the presence of saturated ketone and/or aldehyde
groups [45].   Samples which had undergone photooxidation under
simulated sunshine conditions (Xenotest 150 weathering cabinet for
150 hr) also showed a decrease in emission intensity but with no
shift to longer wavelength.   In fact, samples excited at 310 nm
showed a shift of both emission and excitation maxima to shorter
wavelength.

     The longer wavelength species were consumed when samples of
nylon 6,6 film which had undergone periods of prior thermal oxidation were subjected to photooxidizing conditions.   Table 4 shows how
the intensities of emissions at longer wavelengths decreased relative
to those at 455 nm in the unirradiated film.   Generation of such
thermal oxidation products, as may occur in the processing stage
of the polymer, renders it more photosensitive by increasing its
capacity for absorbing sunlight in the 300- to 350-nm range.

TABLE 4 Effect of Irradiation on the Phosphorescence Emission of Nylon 6,6 Film

Irradiation time (hr)	λ(ex) (nm)	λ(em) (nm)	Lifetime (sec)	Ratio[a]
0	280	455	0.82	1.00
	300	465	0.37	0.95
	320	480	0.21	0.88
	360	500	0.17	0.35
25	280	435	1.20	0.85
	300	460	0.40	0.60
	320	475	0.36	0.38
	360	500	0.18	0.08
68	280	430	1.30	0.80
	300	455	0.45	0.53
	320	470	0.30	0.32
	360	500	0.15	0.05
100	280	430	1.30	0.75
	300	450	0.40	0.50
	320	470	0.27	0.31
	360	500	0.15	0.04

[a]Ratio of emission intensity to that of unirradiated film at 455 nm.

Although the mechanism of the photooxidation of polyamides is still only partially understood, data available so far generally support the conclusions drawn from early work in polyamides as well as model amide molecules that showed that photooxidation favors the formation of nonphosphorescent aliphatic amides and carboxylic acids, which accounts for the absence of long-wavelength phosphorescence emissions. These latter groups which result from thermal oxidation initiate and are consumed by the photooxidative process.

V. POLYURETHANES

Since the appearance of polyurethanes in the 1930s there has been a steady growth in the quantity and range of application of these tough, abrasion-resistant materials based on aromatic and aliphatic diisocyanates. In spite of possessing several advantageous features, however,

the common forms of these materials suffer from an important disadvantage, namely, their sensitivity to ultraviolet light. This sensitivity, which arises from a photooxidative process, manifests itself
as a coloration of the polymer and is accompanied by undesirable
changes in the mechanical properties [50]. The structural cause of
these changes in the case of aromatic systems is thought to be the
oxidation of the diurethane bridge forming monoquinone imide (XII)
and diquinone imide (XIII) structures. The disadvantage imposed by

$$\sim NH\!\!-\!\!\bigcirc\!\!-\!\!CH_2\!\!-\!\!\bigcirc\!\!-\!\!NH\sim \xrightarrow[O_2]{h\nu} \sim NH\!\!-\!\!\bigcirc\!\!-\!\!CH\!\!=\!\!\bigcirc\!\!=\!\!N\sim$$
$$(XII)$$
$$\downarrow$$
$$\sim N\!\!=\!\!\bigcirc\!\!=\!\!C\!\!=\!\!\bigcirc\!\!=\!\!N\sim$$
$$(XIII)$$

(11)

this sensitivity has been minimized by the development of polymers
containing aliphatic structures and by the incorporation of effective
stabilizer systems, but for largely economic reasons polymers based
on aromatic diisocyanates with corresponding diol coreagents still
occupy a significant place in the market and are therefore subjects
of current interest.

Photooxidative fragmentation reactions as well as photo—Fries-type
rearrangements complicate the mechanistic picture very considerably
since it has been suggested that the products may also be precursors
of color products [51]. Beachell and Chang [52] showed that triplet
donors such as benzophenone and anthraquinone sensitize the photooxidation of the model ethyl-N-phenylcarbamate (EPC) by triplet
energy transfer to the triplet state of EPC. Involvement of the triplet state was also deduced by Masilamani et al. [53], but Noack and
Schwetlick [54] concluded on the basis of singlet and triplet quenching studies that the photo—Fries rearrangement of N-phenylcarbamate
originated with the singlet $\pi\pi^*$ state.

Clearly, therefore, emission spectroscopy has an important role
to play in these complicated and diverse systems, and it has been
used to supplement ESR, mass spectrometry, and other spectroscopies.
The emission spectra of polyurethane solutions and film are shown in
Figure 19. The phosphorescence component, which closely resembles
red-shifted benzophenone, has been attributed by Allen and McKeller
[55] to the presence in the polymer backbone of benzophenone-type
"defect" structures (XIV), but the nature of the species responsible
for the fluorescence has not been identified with certainty. During
irradiation under simulated sunlight conditions both fluorescent and

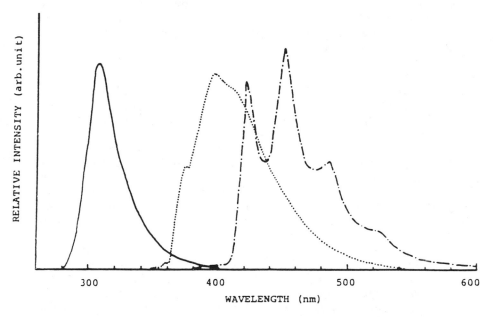

(XIV)

phosphorescent species are consumed. Laser flash photolysis of
diphenylmethane-4,4'-diisocyanate (MDI)-based polyurethane film
produced two transients with absorption maxima at 375 and 580 nm.
As Figure 20 shows, the former transient disappeared relatively
quickly on repeated flashing while the latter, which closely resembled
that of the ketyl radical derived from benzophenone, reacted more
slowly. This behavior has been attributed to a mechanism which
begins with the extraction by the triplet-excited benzophenone defect
in the polymer of a hydrogen atom from the surrounding medium.
The radicals XV to XVII could then take part in a variety of oxidative processes leading to degradation and deterioration of the polymer.

FIGURE 19   Emission spectra of polyurethane: (a) fluorescence emission [λ(ex) = 280 nm] (——) of THF solution; (b) phosphorescence
emission [λ(ex) = 280 nm] (-·-·) of THF solution; and (c) phosphorescence emission [λ(ex) = 310 nm] (-·-·) of film.   (Reproduced from
Polym. Deg. Stab., 1, 311 (1979) with permission from Elsevier Applied
Science Publishers, UK.)

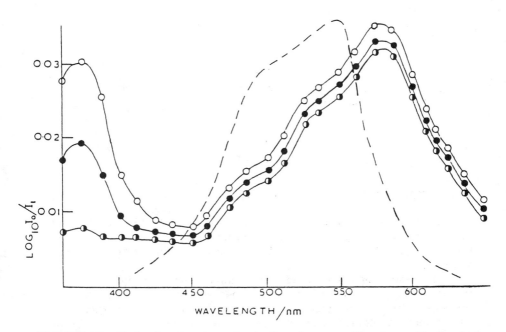

$$\text{(12)}$$

Analogous primary processes have been suggested for the benzophen-one-sensitized oxidation of other polymers including polyolefins [56] and PVC [59].

FIGURE 20 End of pulse transient absorbances after laser flash photolysis of MDI-based polyurethane film (O) one flash, (•) two flashes, and (◑) three flashes compared with alcohol solution of benzophenone ketyl radical (---). (Reproduced from Ref. 55 with permission from John Wiley and Sons, New York.)

VI. POLYETHERSULFONE

Polyethersulfone is mentioned here since it differs from those discussed previously in that the fluorescence and phosphorescence
excitation and emission spectra of the commercial polymer match closely
those of the model diphenylsulfone monomer [58]. This means that
in this case absorption and emission originate within the basic polymer unit and not from an impurity or structural defect as is more
often the case.

The emissions are reduced in intensity as the polymer is irradiated by sunlight under oxidative conditions, and laser flash photolysis
indicates the intermediacy of a species with absorption maxima at
435, 465, and 601 nm. They bear some similarity to phenoxy radicals
derived from flash photolysis of phenols (Figure 21), the positions
of the maxima depending on the degree of substitution of the phenyl
group. The flash photolysis data [58] together with information from
ESR [59] measurements suggest that the degradation is initiated by
scission of the phenyl—sulfone or phenyl—oxygen bond, i.e.:

$$(13)$$

VII. CONCLUSIONS

⟨ab

Luminescence spectroscopy has proved to be an invaluable adjunct
to the absorption spectrographic techniques which have been used
to investigate the mechanisms of photooxidative processes in polymer
systems. In the paper
⟨c⟩ —Attention has centered on three main areas, namely, (1) the
identification of absorbing chromophores which initiate the degradation
process, (2) characterization of oxygenated and other photoproducts
which cause a deterioration in the polymer properties, and (3) quantification of the energy transfer process which in some cases make the
sensitized process dominant. In all cases the high sensitivity and
selectivity of the technique has proved invaluable when the species
under investigation are present at very low concentration levels and

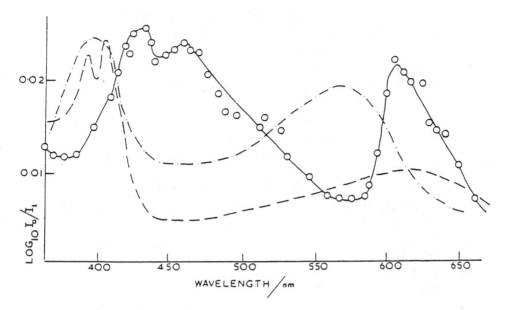

FIGURE 21   End-of-pulse transient absorbance after laser flash photolysis of poly(ethersulfone) film (0) compared with that of a phenoxy
radical (∅0°) in paraffin (---) and a triphenoxy radical in 50% ethanol
(-·-·). (Reproduced from Ref. 58 with permission from John Wiley
and Sons, New York.)

the information gained has led to important indications about the way
in which polymers can be effectively protected (stabilized) against
the harmful effects of sunlight.

   Although mechanistic details are complex and diverse, many
similarities have emerged as detailed data from different systems have
become available. There is no doubt that the rapidity and simplicity
with which measurements can be made will ensure that luminescence
spectroscopy will continue to be a routine technique in polymer
photooxidation studies. *Graphs, spectra.*

REFERENCES

1. A. C. Somersall and J. E. Guillet, J. Macromol. Sci., Rev.
    Macromol. Chem., C(B): 135 (1975).

2. A. Garton, D. J. Carlson, and D. M. Wiles, Develop. Polym.
    Phot., 1: 93 (1980)

3. K. B. Chakraborty and G. Scott, Eur. Polym. J., 13: 731 (1977)

4. G. Scott, Am. Chem. Soc. Symp. Ser., 25: 340 (1976).

5. D. J. Carlson and D. M. Wiles, Macromolecules, 2: 587 (1969).

6. F. Sitek, J. E. Guillet, and M. Heskins, J. Polym. Sci., Symp. Ser., 57: 343 (1976).

7. N. S. Allen, J. Homer, and J. F. McKeller, J. Appl. Polym. Sci., 20: 2553 (1976).

8. N. S. Allen, J. Homer, and J. K. McKeller, J. Appl. Polym. Sci., 21: 2261 (1977).

9. A. Charlesby and R. H. Partridge, Proc. Roy. Soc., A283: 312 (1965).

10. S. W. Beavan and D. Phillips, J. Photochem., 3: 349 (1974).

11. D. Bellis, D. R. Kearns, and K. S. Schaffner, Helv. Chim. Acta, 52: 971 (1976).

12. D. J. Carlson and D. M. Wiles, J. Polym. Sci., Polym. Letts. Ed., 11: 759 (1973).

13. D. J. Carlson, T. Suprunchuk, and D. M. Wiles, J. Polym. Sci., B-11: 149 (1973).

14. J. P. Dalle, R. Magous, and M. Mouserron-Canet, Photochem. Photobiol., 15: 411 (1972).

15. Z. Osawa and H. Hiroda, J. Polym. Sci., Polym. Letts. Ed., 20: 577 (1982).

16. N. S. Allen, J. F. McKeller, and G. M. Wood. J. Polym. Sci., Polym. Chem. Ed., 13: 2319 (1975).

17. S. W. Beavan and D. Phillips, Eur. Polym. J., 10: 593 (1974).

18. N. Grassie and N. A. Weir, J. Appl. Polym. Sci., 9: 963, 975, 987, 999 (1965).

19. G. A. George and D. K. C. Hodgeman, J. Polym. Sci., Symp. Ser., 55: 195 (1976).

20. J. B. Lawrence and N. A. Weir, J. Polym. Sci., Polym. Chem. Ed., 11: 105 (1973).

21. G. Geuskens, D. Baeyens-Volant, G. Delaunois, Q. Lu-Vinh, W. Piret, and C. David, Eur. Polym. J., 14: 291 (1978).

22. G. A. George and D. K. C. Hodgeman, Eur. Polym. J., 13: 63 (1976).

23. H. H. Jaffe and M. Orchin, Theory and Applications of Ultraviolet Spectroscopy, John Wiley and Sons, New York (1962).

24. L. J. Basile, Trans. Faraday Soc., 60: 1702 (1964).

25. J. B. Berlman, Handbook of Fluorescence Spectra of Aromatic Molecules, 2nd ed., Academic Press, New York (1971).

26. G. Geuskens, D. Baeyens-Volant, G. Delaunois, Q. Lu-Vinh, W. Piret, and C. David, Eur. Polym. J., 14: 299 (1978).

27. E. D. Owen, Dev. Pol. Photochem., 3: 165 (1982).

28. C. Decker, Degradation and Stabilization of PVC (E. D. Owen, ed.), Elsevier, London, p. 81 (1984).

29. R. J. Bailey, Ph.D. thesis, University of Wales (1972).

30. E. D. Owen and R. L. Read, Eur. Polym. J., 15: 41 (1979).

31. F. Sondheimer, D. A. Ben-Efrain, and R. Wolovsky, J. Am. Chem. Soc., 83: 1675 (1961).

32. G. Palma and M. Carenza, J. Appl. Polym. Sci., 16: 2485 (1972).

33. W. I. Bengough and T. K. Varma, Eur. Polym. J., 2: 61 (1966).

34. Y. Shindo and T. Hirai, Makromol. Chem., 155: 1 (1972).

35. Y. Shindo, B. E. Read, and R. S. Stein, Makromol. Chem., 118: 272 (1968).

36. M. Balandier and C. Decker, Eur. Polym. J., 14: 995 (1978).

37. P. K. Das and R. S. Becker, Photochem. Photobiol., 32: 739 (1980).

38. R. S. Becker, R. V. Benassen, J. Lafferty, T. G. Truscott, and E. J. Land, J. Chem. Soc., Faraday Trans.II, 74: 2246 (1978).

39. P. K. Das and R. S. Becker, J. Phys. Chem., 86: 921 (1982).

40. K. Maruyama, H. Akahoshi, M. Kobayoshi, and Y. Tanizaki, Bull. Chem. Soc. Japan, 58: 2923 (1985).

41. K. Maruyama, H. Akahoshi, M. Kobayashi, and Y. Tanizaki, Chem. Lett., 1863 (1983).

42. T. Hjertberg and E. M. Sorvik, Degradation and Stabilization of PVC (E. D. Owen, ed.), Elsevier, London, p. 21 (1984).

43. N. S. Allen, J. F. McKeller, and D. Wilson, J. Polym. Sci., Polym. Chem. Ed., 15: 2793 (1977).

44. N. S. Allen, J. F. McKeller, and G. O. Phillips, J. Polym. Sci., Polym. Chem. Ed., 12: 2623 (1974).

45. N. S. Allen, J. F. McKeller, and D. Wilson, J. Photochem., 6: 337 (1976).

46. H. S. Koenig and C. W. Roberts, J. Appl. Polym. Sci., 19: 1847 (1975).

47. J. A. Dellinger and C. W. Roberts, J. Appl. Polym. Sci., Polym. Letts. Ed., 14: 167 (1976).

48. D. B. Larson, J. F. Arnatt, and S. P. McGlynn, J. Am. Chem. Soc., 95: 6928 (1973).

49. D. B. Larson, J. F. Arnatt, C. J. Seliskar, and S. P. McGlynn, J. Am. Chem. Soc., 96: 3370 (1974).

50. C. S. Schollenberger and F. D. Stewart, in Advances in Polyurethane Science and Technology, Vol. 2 (K. C. Frisch and S. L. Reegen, eds.), Technomatic, p. 71 (1973).

51. Z. Osawa, Dev. Polym. Phot., 3: 209 (1982).

52. H. C. Beachell and I. L. Chang, J. Polym. Sci., A-1, 10: 503 (1972).

53. D. Masilamani, R. O. Hutchins, and J. Ohr, J. Org. Chem., 41: 3687 (1976).

54. R. Noack and K. Schwetlick, Tetrahedron, 30: 3799 (1974).

55. N. S. Allen and J. F. McKeller, J. Appl. Polym. Sci., 20: 1441 (1976).

56. D. J. Harper and J. F. McKeller, J. Appl. Polym. Sci., 17: 3503 (1973).

57. E. D. Owen and R. J. Bailey, J. Polym. Sci., A-1, 10: 113 (1972).

58. N. S. Allen and J. F. McKeller, J. Appl. Polym. Sci., 21: 1129 (1977).

59. B. D. Gesner and P. G. Kelleher, J. Appl. Polym. Sci., 13: 2183 (1969).

6

Stress-Induced Chemiluminescence

SUZANNE B. MONACO and JEFFERY H. RICHARDSON   Lawrence
Livermore National Laboratory, Livermore, California

Lab. p257

I. INTRODUCTION

A nondestructive, nonhysteretic technique that could predict the
probability of failure of a stressed polymeric material in a given environment would provide a valuable quality control test.   Such a
technique would narrow the distribution function of failure probability
vs. stress during various processing steps.   Other technical applications would be in assessing long-term durability of the material under
expected service conditions.   Polymeric materials find ever increasing
usage in industrial and commercial applications in a wide range of
service conditions.   Many of these industrial applications make use
of the favorable strength-to-weight characteristics of polymeric materials (e.g., fiber composites).   However, little is known about the
long-term durability of these polymeric materials; any deleterious
aging reactions are very slow, requiring years to cause component
failure.   Many of these materials (particularly fiber composites) are
stressed during their service life causing these aging reactions to
change or, at a minimum, to increase in rate [1].   Thus, there is
considerable interest in a technique of sufficient sensitivity to predict the performance of materials as a function of both processing

*Work performed under the auspices of the U.S. Department of
Energy by the Lawrence Livermore National Laboratory under con-
tract number W-7405-ENG-48.

237

parameters and parameters characteristic of the intended service
(e.g., time, stress, atmosphere, temperature). Besides serving as
an empirical data base, such technical information could also provide
insight into the role of stress-induced reactions in the aging process
of polymeric materials.

Many groups have looked at the effect of stress on polymer-
aging reactions using techniques such as IR [2], FTIR [3,4], Raman
[5], EPR [6], and mass spectroscopy [7]. These studies have yielded mechanistic information but lack the sensitivity needed for a nondestructive quality control test of processing variables or for evaluation of long-term durability under typical service environments such
as constant or variable stress.

In order to gain enhanced sensitivity, polymeric materials are frequently studied by conventional chemiluminescence techniques [8,9].
This highly sensitive technique can detect degradation changes due
to aging and temperature in various polymeric materials (e.g., resins,
fibers, rubbers, food products). This enhanced sensitivity permits
the detection of these microscopic degradation reactions up to tens
of years before their effect would become apparent on a macroscopic
scale. Since mechanical damage to some of these polymeric materials
results in bond scission and free-radical reactions, it was a logical
step to look for the chemiluminescence of polymeric materials in the
presence of an oxidant and while the material was under stress.
Stress chemiluminenscence (SCL) is a term coined by Fanter and Levy
[10,11] and refers to just such a process.

The initiation step in SCL is bond scission from mechanical stress
(as opposed to heating or chemical reactions in ordinary chemiluminescence). After the initial step, the rest of the mechanism is assumed
to be similar to conventional polymeric chemiluminescence: subsequent
reactions with oxygen and a procession through peroxide intermediates give rise to electronically excited species which luminesce
(e.g., ketones). The work in the literature to date has dealt primarily with SCL from crosslinked resins, oriented fibers, and fiber
composites. Correlations have been made of chemiluminescence data
with applied stress, temperature, atmosphere, absorbed moisture,
and accelerated aging for samples of some widely different polymeric
materials. In addition, there are some data correlating an enhanced
chemiluminescence signal in a low-stress environment with subsequent
premature failure of the polymeric material. These results are currently not firmly established for a very wide variety of materials and
may not be widely applicable. Furthermore, like any chemiluminescence technique, stress chemiluminescence is subject to the usual
problems of self-quenching, self-absorption, surface effects, and
trace impurities. Thus, at the present time stress chemiluminescence
presents an intriguing but as yet unfulfilled potential for predictive,
nondestructive testing of polymeric materials.

II. THEORY

There are, of course, a vast number of mechanisms which result in the emission of light; many different substances produce light under a variety of circumstances. Chemiluminescence and triboluminescence are two of the more common processes which result in the emission of light. Stress chemiluminescence is a subset of chemiluminescence and needs to be distinguished from triboluminescence.

Chemiluminescence of polymeric materials is generally perceived as proceeding through a free-radical mechanism involving primary or secondary peroxy radicals, culminating in the production of electronically excited carbonyl groups [8,9]. It is the radiative decay of these carbonyl groups which is responsible for the light emission. The initiation step involves the reaction of the polymeric material with oxygen (or an oxidizing agent) under conditions of heat, radiation, and/or specific catalysts. The detailed free-radical mechanisms responsible for the chemiluminescence of polymeric materials are described elsewhere in this volume.

Triboluminescence is defined as "the emission of light caused by the application of mechanical energy to a solid" [12]. There are many examples of triboluminescence, including those involving polymeric materials [12,13]. Triboluminescence takes its name from the Greek tribein, to rub. Chemiluminescence, on the other hand, is light emission caused by chemical reaction and is often referred to as "cold light." Chemiluminescence of polymeric materials is interpreted in terms of exothermic oxidation reactions and thus cannot occur in the absence of an oxidant, whereas triboluminescence can. Stress chemiluminescence is interpreted as being initiated by mechanical deformation of bonds, but the subsequent radical chain and eventual emission of light is similar to that associated with chemiluminescence. Consequently, stress chemiluminescence does not occur in the absence of an oxidant, either. In some circumstances both stress chemiluminescence and triboluminescence can occur simultaneously [14]. There have been several reported examples of stress chemiluminescence where the dependence on both oxygen and stress have been explicitly demonstrated [10,11,15−21].

III. EXPERIMENTAL PROCEDURES

A. Apparatus

The basic equipment needed to study stress chemiluminescence is actually relatively simple. All that is required is a light-tight environmental chamber with temperature and atmospheric control, provisions for applying and measuring stress (and ideally extension) to a sample of polymeric material, a sensitive light detector, and signal-processing electronics and recording equipment. In addition, the

capability for doing wavelength discrimination is desirable. Of the
three systems reported in the literature [10,16,18], one consists of
an Instron tensile testing machine modified with an environmental
chamber and light collection system [16], and the other two were designed and built especially for stress chemiluminescence measurement
[10,18,22]. These two also have provisions for controlling the atmosphere surrounding the sample since their environmental chambers
are considerably smaller than that of the modified tensile testing
machine.

A schematic of the system used by the authors, which is fairly
representative of the others, is depicted in Figure 1 [18] and will
be described in some detail. An LSI-11 microprocessor is used to
ramp a stepping motor which applies stress to the sample of polymeric
material (up to 70 MPa for samples 0.2 cm^2). Various gripping mech-

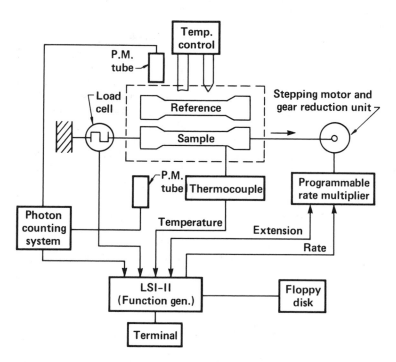

FIGURE 1 Schematic of a typical stress chemiluminescence apparatus.
A dual-channel photon-counting system monitors the luminescence as
a function of wavelength, stress, temperature, atmosphere, and elongation. The entire apparatus is under computer control, for both
data acquisition and reduction. (Reprinted with permission from
Ref.18. Copyright 1982 American Chemical Society.)

anisms are used depending on the nature of the polymer sample
(e.g., dog bone specimens, fibers, elastomers). The computer also
reads the output of the load cell and servos the stepping motor
until the desired stress level is reached. A software feedback loop
between the load cell and motor permits the programming of specific
stress-time sequences. Various load cells are used with maximum
readouts ranging from 25 to 500 lb depending on the type of sample.
In the case of elastomers, which can elongate beyond 400%, extension
rather than stress is measured and controlled by the computer. A
separate unit controls the temperature of a heater placed beneath
the sample. The temperature of the sample is read by the computer
via a thermocouple placed between the sample and heater. The maximum temperature achievable is 170°C. The light-tight sample chamber can be evacuated and then filled with the atmosphere of interest
(e.g., air, oxygen, nitrogen).

Chemiluminescence is detected with two photomultiplier tubes
(RCA 8852), typically biased to 2000 V. The tubes are kept at
-30°C with thermoelectric refrigerated chambers which are water-
cooled. Dark counts of the tubes are typically 300 counts/min.
One tube is placed opposite the sample to be stressed and the other
opposite the unstressed blank or reference sample. The chemiluminescence of each is imaged on the 2-in.-diameter photocathode by a
50-mm f/0.95 camera lens and an antireflection coated plano-convex
lens with a focal length of 90 mm. Outputs from each tube are separately routed through PAR 1121A amplifier discriminators to the
PAR 1112 photon counter/processor and simultaneously stored in two
channels. The contents of each channel or the difference between
the two can be displayed and sent to the computer. When operating
in the difference mode, the chemiluminescence signal of an identical
sample exposed to the same temperature and atmosphere is continually
subtracted from the chemiluminescence signal of the stressed sample
of polymeric material. Provisions exist for wavelength discrimination
using long-pass filters; work by Fanter and Levy [10,11] showed
that stress chemiluminescence exhibited a shorter wavelength than
chemiluminescence initiated by thermal heating. The LSI-11 outputs
the stress level (or extension), temperature, and corresponding
photon counts to both a printer and a floppy disk for storage and
subsequent data manipulation.

B. Materials

The polymeric materials studied to date with stress chemiluminescence
have been primarily crosslinked resins, oriented fibers, and fiber
composites because of the emphasis on developing a nondestructive
quality control test for fiber composite components. George et al.
studied nylon 66 warp yarns and plain woven fabric from samples
of widely different weight and yarn properties [16]. Fanter and

coworkers at McDonnell Douglas Corporation did the bulk of their studies with tetraglycidyl 4, 4'-methylenedianiline (TGMDA or TGDDM) cured with 4, 4'-diaminodiphenylsulfone (DDS), the epoxy resin most commonly used in high-performance fiber composites [10,11, 20—23]. Both commercial formulations and specially prepared samples containing varying amounts of crosslinking agent were used. The special samples were prepared from components purchased from Ciba-Geigy (MY720 and Eporal) and were prepared by a method based on curing the resin mixture in silicone rubber molds [24]. Most of the authors' work was also done with samples of TGMDA-DDS formulated with the materials from Ciba-Geigy and without catalyst. However, our specimens were cut and polished from resin cured in a sheet between glass plates [17—19]. Care was taken in the handling and the samples were stored in the dark prior to testing. The application of fiberglass end tabs was found to greatly increase the reliability of the gripping mechanism.

As epoxy elongates less than 2%, the area viewed by the photo-multiplier tube was essentially constant.

Some stress chemiluminescence work with silicone cushions has also been reported [18]. Silicone cushions are made from silicone gum, SiO_2 fillers, and peroxide curing agents. Urea functions as a temporary pore former. The exact formulations as well as an investigation of accelerated aging by chemiluminescence has been described [18,25]. Samples using elastomers were always physically wide enough that any narrowing from elongation did not reduce the area viewed by the photomultiplier tube.

IV.   RESULTS AND DISCUSSION

Relatively few results of stress-induced chemiluminescence have been reported in the literature. This section summarizes the salient observations and indicates the general nature of the information which can be gleaned from investigations of polymeric materials by stress-induced chemiluminescence techniques.

One of the earliest reports of stress-induced chemiluminescence was of polymeric elastomers by Butyagin and coworkers [14]. This qualitative work was an investigation of the deformation of polymer films by air pressure and the concomitant measurement of total luminescence. The polymeric materials studied were natural rubber, polyethylene, polypropylene, ethylene-propylene copolymers, ethylene-butylene copolymers, polyethyleneterephthalate, and polytetrafluoroethylene. With all these films it was observed that luminescence occurs when the films were stretched, were suddenly contracted, or upon rupture of the films. In general, the luminescence intensity increased with both rate of deformation and film thickness. The maximum luminescence intensity correlated with the percent elongation at the threshold of luminescence detection. Additives significantly affect the luminescence intensity (e.g., unplasticized polyvinylchloride vs. material plasticized with dioctyl sebacate). The mechanism is discussed in terms of chemiluminescence, but no specific data are presented to distinguish the luminescence from triboluminescence.

A major contribution in the field of stress-induced chemiluminescence was the work of Fanter, Levy, and coworkers [4,10,11,20–24]. This work represented the first major correlation of luminescence intensity with applied stress for polymeric materials. Most of the polymeric materials studied by Fanter and Levy were epoxy resins (TGDDM/DDS). Figure 2 illustrates a typical correlation between applied stress and observed chemiluminescence intensity for an epoxy resin. Furthermore, triboluminescence was eliminated as an explanation for this luminescence by a series of experiments in air, nitrogen, and oxygen. Figure 3 is representative of the experimental data indicating that this luminescence is indeed chemiluminescence, and is dependent on the presence of an oxidizing agent (in this case oxygen).

Fanter and Levy speculate about the mechanism of the stress-induced chemiluminescence signal but do not have experimental evidence to distinguish which (if any) of several suggested mechanisms are actually dominant in the luminescence process. The mechanical stress on the polymeric material is suggested to have two possible effects:

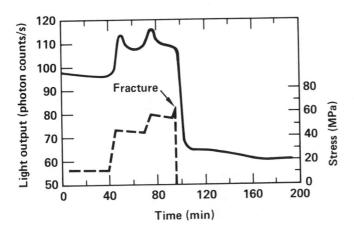

FIGURE 2 Stress chemiluminescence intensity of epoxy resin (TGDDM/DDS) at 109°C and in an oxygen atmosphere. Also shown is applied stress to the sample; the correlation between luminescence intensity and stress is apparent. (Reprinted with permission from Ref. 20. Copyright 1979 American Chemical Society.)

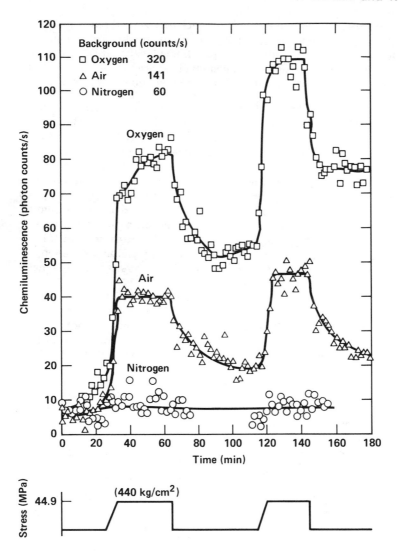

FIGURE 3 Chemiluminescence of epoxy resin materials (TGDDM/DDS) as a function of both stress and atmosphere. The absent luminescence intensity in the absence of oxygen is indicative of a chemiluminescence mechanism as opposed to triboluminescence. (Reprinted with permission from Ref. 21. Copyright 1979 American Chemical Society.)

(1) Mechanical stress applied to the polymer chains may produce an increase in both the intramolecular distance (i.e., chain deformation) and interatomic distance (i.e., bond deformation) without causing bond scission. Such an alteration in the local environment may

result in a change in the local potential energy configuration, leading to the enhancement of luminescence reactions between the polymer chain and surrounding molecules (e.g., oxygen) with or without free-radical formation.

(2) At some point mechanical stress applied to the polymeric material results in bond scission and accompanying formation of free radicals, along with a decrease in the activation energy for thermally induced oxidation reactions. At this point reaction with molecular species in the local environment is similar to that of thermally induced reactions resulting in chemiluminescence. However, Figure 4 illustrates the difference in the spectral distribution of the chemiluminescence of stressed and unstressed epoxy resin (TGDDM/DDS) in oxygen. While this shift in spectral distribution is not always seen, its occurrence suggests either that the stressed polymer can access reactions less favorable (and characterized with different luminescence spectra) for unstressed polymers or that the distribution of excited states for the stressed sample is different and more extensive than that for the unstressed polymer. (The latter appears less likely considering the relaxation times for excited species in the solid state.)

Fanter and Levy invariably observed a change in luminescence intensity as a function of time for a constant applied stress. This

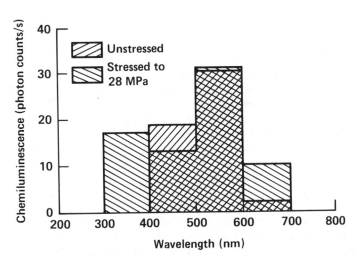

FIGURE 4 Spectral distribution of stressed and unstressed epoxy resin (TGDDM/DDS) in oxygen at 140°C. This difference in luminescence spectra suggests some difference in the excited state population distribution or the accessibility of luminescence reactions. (Reprinted with permission from Ref. 20. Copyright 1978 American Chemical Society.)

observation remains unexplained, but may be attributed to several possible factors: (1) creep of the material and the associated change in optical viewing area or accessibility of oxygen to the material; (2) if the stress-induced chemiluminescence signal only originates from the surface, as the stress-activated bonds react there will be an accompanying change in the luminescence signal; (3) the chemiluminescence signal also originates from the bulk of the material, and hence after an initial period the reaction is dominated by oxygen diffusion. A counterpoint to the last suggested reason was the observation by Butyagin [14] that the luminescence signal (for a different class of polymeric materials) was linear in thickness.

Fanter and Levy also illustrated the dependence of the stress-induced chemiluminescence signal on both the composition of the polymeric material and the environment. Figure 5 illustrates three different formulations of a TGDDM/DDS epoxy resin: 23 phr DDS (without catalyst), 27 phr DDS (without catalyst), and 24 phr DDS with a propretary catalyst. The former two are very similar (identical within the reproducibility of the experiment), but the material with the catalyst is significantly different either reflecting a structural difference or the additive effect of the catalyst (observed by Butyagin [14] but at much higher additive levels and in a different class of polymeric materials).

Figure 6 illustrates the effect of moisture on the stress chemiluminescence signal. This result is certainly preliminary but it illustrates the potential of stress chemiluminescence for monitoring changes in properties of polymeric materials which are a function of the environment (in this case humidity).

The authors' interest in stress chemiluminescence was motivated by an interest in predicting the probability of failure of polymeric materials in a given environment without destroying the material (i.e., a technique with a nondestructive predictive capability). The results of a preliminary investigation of stress-induced chemiluminescence as such a technique were described in several papers [17–19, 26] and will be summarized here.

All subsequent data represented here [17–19,26] from the various materials examined (epoxy resins, Kevlar fibers, silicone cushions) were taken after allowing an initial large signal to decay to a much lower and more slowly varying value. The source of this initial signal is not clearly understood but may be due, at least in part, to photooxidative degradation resulting from exposure to ambient light. Decay times were approximately 1–2 hr and did not follow a single exponential. They did, however, follow the linear relation described by George [16]. Care was taken to keep samples in the dark, which significantly reduced this initial signal.

Figures 7 and 8 illustrate the dependence of the luminescence signal on stress and oxygen (hence stress-induced chemiluminescence). Unlike Levy and Fanter (e.g., Figure 5), our results in-

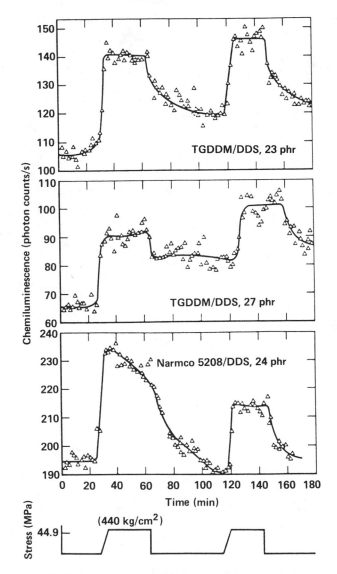

FIGURE 5 Chemiluminescence signal of three different formulations of epoxy resin (TGDDM/DDS) in air and at room temperature as a function of applied stress: 23 phr DDS (no catalyst), 27 phr DDS (no catalyst), and 24 phr DDS (with proprietary catalyst). The first two are essentially identical within experimental error but the third clearly is different, reflecting either structural or additive effects. (Reprinted with permission from Ref. 21. Copyright 1979 American Chemical Society.)

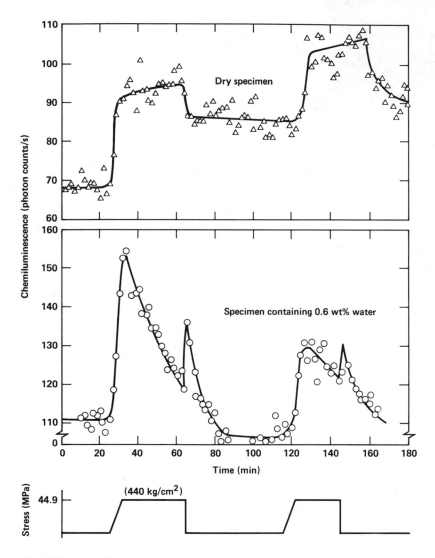

FIGURE 6   Effect of moisture on the stress chemiluminescence signal
of epoxy resin (TGDDM/DDS).   These data illustrate the potential
of stress chemiluminescence for monitoring polymeric properties as a
function of the environment.   (Reprinted with permission from Ref.
21.   Copyright 1979 American Chemical Society.)

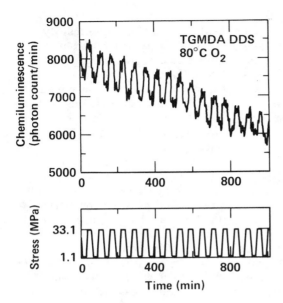

FIGURE 7   Stress-induced chemiluminescence signal of epoxy resin
sample (TGMDA/DDS) in oxygen at 80°C.   (Reprinted with permission from Ref. 18.   Copyright 1982 American Chemical Society.)

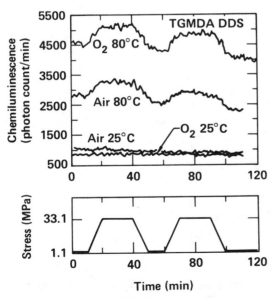

FIGURE 8   Stress-induced chemiluminescence of epoxy resin sample
(TGMDA/DDS) as a function of temperature and atmosphere.   (Re-
printed with permission from Ref. 18.   Copyright 1982 American
Chemical Society.)

variably showed a dropping baseline as opposed to a rising baseline. Data taken at 80°C in nitrogen did not differ significantly
from data shown in Figure 8 taken at room temperature and in air;
consequently, the stress-induced luminescence signal is not triboluminescence and is certainly related to oxidative reactions.

Since absorbed moisture lowers the glass transition temperature
of the TGDDM/DDS epoxy resin system and has a deleterious effect
on the performance of the material, the effect of water on the stress-
induced chemiluminescence intensity was examined. Figure 9 illustrates the stress-induced chemiluminescence at 80°C in oxygen of
the same epoxy resin sample before and after the introduction of
water. The sample was exposed to 100% relative humidity at room
temperature for 24 hr and experienced a weight gain of 0.3%. Samples that absorbed more than 0.6% water broke at such low stress
that no data were obtained. The data illustrated in Figure 9 clearly
show the following: (1) As was the case with Fanter and Levy
(Figure 6), adsorbed moisture has a major effect on the stress-
induced chemiluminescence signal of epoxy resin materials. (2)
The direction and shape of the stress-induced chemiluminescence
signal is different in the two cases. The exact reasons for this
difference are not known, but are likely due to differences in for-

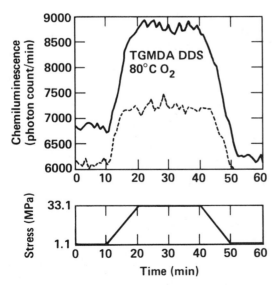

FIGURE 9 Stress-induced chemiluminescence of epoxy resin sample
(TGMDA/DDS) at 80°C in oxygen before (solid line) and after (broken line) introduction of 0.3 wt % water. (Reprinted with permission from Ref. 18. Copyright 1982 American Chemical Society.)

mulation and preparation of the sample, as well as the mechanism of water introduction.

The predictive nature of stress-induced chemiluminescence was investigated by preparing a series of samples in identical fashion, and then dividing the samples into two groups. One group functioned as a control, and the other experienced various postfabrication treatments. In all cases the stress-induced chemiluminescence test was the same.

Four samples were stressed to 33.1 MPa at 80°C in oxygen and their stress-induced chemiluminescence intensity recorded. Figure 10a is typical of the data taken. All of the samples exhibited a 10 ± 2% increase in luminescence at this stress level and subsequently failed at loads greater than 48 MPa. These samples were taken to be of high quality and their chemiluminescence under these conditions formed a basis for comparison with lower quality and deliberately damaged samples. Table 1 lists the change in the chemiluminescence signal with stress for four epoxy samples (b, c, d, and e), which ultimately were determined to be of poor quality on the basis of a lower tensile strength. These results indicate a potential predictive capability of the stress-induced chemiluminescence technique. Sample a represents the four samples that exhibited a high tensile strength. Samples b and e were damaged by scoring with a scalpel in the narrow portion of the dog bone. They showed a 15 and 19.6% change in stress-induced chemiluminescence intensity and failed at 43.9 and 31.7 MPa, respectively. Both samples failed at the score. Sample c was placed in an oven at 70°C for 1 week. A 21% increase in stress-induced chemiluminescence intensity was seen and the sample failed at 43.8 MPa. It is clear that the effects of accelerated aging can be detected by stress-induced chemiluminescence. Sample d was not deliberately damaged. An increase of 19.5% was seen in the stress-induced chemiluminescence intensity and the sample failed at 43.1 MPa. The reason for failure is unknown but could relate to formulation or sample preparation.

The encouraging result is a qualitative correlation on a very limited statistical sampling of an enhanced chemiluminescence signal at low levels of applied stress with the premature mechanical failure of the polymeric material samples. For example, sample e showed an elevated chemiluminescence signal at 12 MPa (Figure 10b). This result was reproducible; in no instance did a sample with a high tensile strength result in a stress-induced chemiluminescence signal at this low value of applied stress. These results, while clearly limited in number, do suggest that stress-induced chemiluminescence has a predictive capability of premature mechanical failure for polymeric materials.

Both Fanter and Levy [10] as well as Monaco and Richardson [19,26] reported stress-induced chemiluminescence from fibers (nylon 66 and Kevlar, respectively). A more thorough investigation of

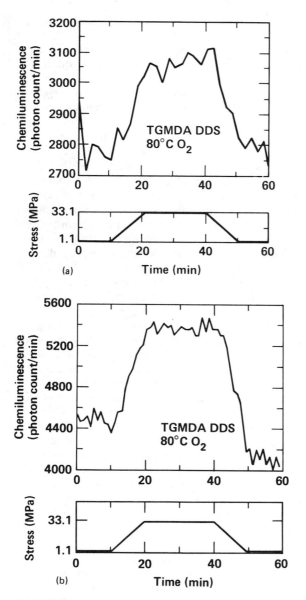

FIGURE 10 Stress-induced chemiluminescence of epoxy resin samples (TGMDA/DDS) at 80°C in oxygen. (a) Sample failed at 48
MPa; (b) sample failed at 31.7 MPa. (Reprinted with permission
from Ref. 18. Copyright 1982 American Chemical Society.)

TABLE 1   Stress-Induced Chemiluminescence of Polymer Characteristics (Epoxy Resin, TGMDA/DDS).[a]

Sample	Characteristics	ΔSCL (%)	Load at failure (MPa)
a	Untreated	10±2	>48
b	Scored	15	43.9
c	Accelerated aged	21	43.8
d	Untreated	19	43.1
e	Scored	20	31.7

[a]All data were taken at 80°C in oxygen with an identical load ramp
to 33.1 MPa.

stress-induced chemiluminescence from nylon 66 fibers was reported
by George and coworkers [9,16].   Once again the stress-induced
luminescence was showed to be chemiluminescence as opposed to
triboluminescence by its dependence on oxygen and absence in nitrogen [16].   With this polymeric material (unlike epoxy resin samples)
there was no apparent wavelength shift between thermally induced
chemiluminescence and stress-induced chemiluminescence; the spectral distributions were similar and the effect of stress was mainly to
enhance the chemiluminescence signal.

Figure 11 illustrates the dependence of the stress-induced
chemiluminescence signal on the rate of applied strain, a parameter
not previously investigated.   Figure 12 illustrates the effect of a
cyclic load on the stress-induced chemiluminescence signal with
nylon 66; it is apparent that a large amount of luminescence signal
(an exponential component) is not present until the previous load
level experienced is exceeded.   Consequently, it is possible that
stress-induced chemiluminescence would be of potential use in the
investigation of the load history of fibers in a given stress-strain
environment.   This observation is in contrast to what was observed
with Kevlar fibers subjected to cyclic loading (Figure 13) [19,26].
However, in each case (Kevlar and nylon 66) a linear or nearly linear Arrhenius relation was observed for the temperature dependence
of the stress-induced chemiluminescence.

The stress-induced chemiluminescence of nylon 66 was interpreted in terms of both a weak component and a strong component
[16]; the former is linear in stress and the latter is exponential in
stress.   The exponential component is interpreted in terms of bond
scission.   This interpretation is at least qualitatively similar to that
advanced for EPR studies of free-radical formation in nylon 66 under
stress [16,27].   Table 2 lists the calculated relative rates of free-
radical formation in nylon 66 with applied stress at 40°C from the

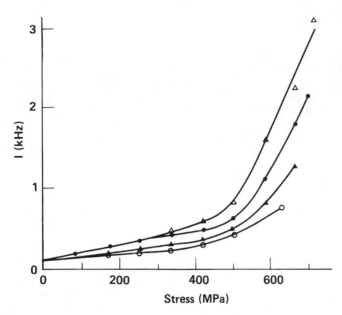

FIGURE 11 The dependence of the stress-induced chemiluminescence signal from high-tenacity nylon 66 yarns at 80°C on the
strain rate. $d\underline{e}/dt$ (min^{-1}): (Δ) 0.25; (·) 0.1; (▲) 0.025; (o)
0.0063. (Reprinted with permission from Ref. 16. Copyright ©
1982 John Wiley and Sons, Inc.)

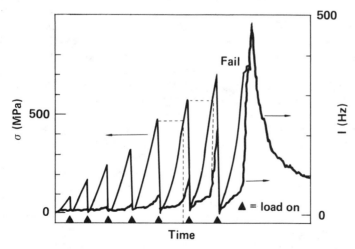

FIGURE 12 The change in stress-induced chemiluminescence intensity from high-tenacity nylon 66 yarns during cyclic incremental
loading at 60°C. There is a time delay between each load/unload
cycle to allow the luminescence signal to decay. σ is the fiber
stress. (Reprinted with permission from Ref. 16. Copyright ©
1982 John Wiley and Sons, Inc.)

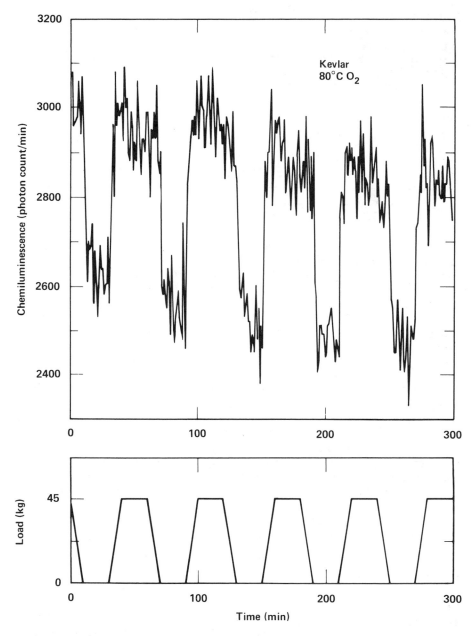

FIGURE 13   Stress-induced chemiluminescence signal from Kevlar fibers in oxygen at 80°C subjected to cyclic loading.   (Reprinted with permission from Ref. 19.)

TABLE 2 Relative Rate of Free-Radical
Formation in Nylon 66 at 40°C with
Applied Stress

Stress (MPa)	σ/σ_f	Relative rate
0	0.00	1.0
150	0.17	2.3
300	0.34	3.4
460	0.52	5.0
715	0.80	7.2

Source: Reprinted from Ref. 16. Copyright © 1982 John Wiley and Sons, Inc.

stress-induced chemiluminescence measurements [9]. The observed
enhanced chemiluminescence signal at low stress implies that damage
to the fibers can occur at low load levels. The linear component
is interpreted in terms of both viscoelastic dissipation of energy
and interfilament friction. Qualitatively the decay of the luminescence after fiber failure, during fiber loading, and after fiber unloading is interpreted in terms of the various possible fates of the
stress-generated alkyl peroxy macroradicals. A slower decay is observed when the fibers remain loaded, which is consistent with an
increase in the free volume of the strained polymer and hence an
increased probability of escape for the radicals from the initiation
molecular cage. Thus, the results of George and coworkers can be
interpreted in terms of the local molecular environment, and hence
stress-induced chemiluminescence may be a probe on the molecular
scale of the effect of macroscopic mechanical deformation. This is
another example of the use of chemiluminescence techniques to study
processes on the microscopic molecular scale and relate the observations to macroscopic processes. A similar correlation was made during a chemiluminescence investigation of compression set and accelerated aging in silicone cushion materials [25].

V. CONCLUSIONS

Stress-induced chemiluminescence is a relatively new technique for
studying polymeric materials. While not ubiquitous, it does appear
to be applicable to a wide variety of different polymer systems.
There is little reported work in the literature on stress-induced
chemiluminescence, and what does exist is very qualitative in nature. However, already there is ample evidence to separate stress-

induced chemiluminescence from triboluminescence as a distinct microscopic process. Furthermore, work with two types of polymeric materials, epoxy resins and nylon fibers, suggests that stress-induced chemiluminescence can be used to probe microscopic processes and interpret the results relative to macroscopic deformations. In each case the effect of microscopic damage to the material was detected by an enhanced chemiluminescence signal well before macroscopic failure of the material occurred. Consequently, stress-induced chemiluminescence may find applications as a nondestructive, nonhysteretic technique for predicting the probability of failure of a stressed polymeric material in a given environment. Important environmental factors have been shown to be temperature, atmosphere, humidity, and current as well as historical stress, strain, and/or strain rate. Important material parameters have been shown to be composition, presence of trace components such as plasticizers and antioxidants, and macroscopic mechanical damage. Such a sensitive tool would be of enormous benefit as a quality control test and as an aid in predicting and monitoring useful service lifetimes of polymeric materials. *Graphs.*

REFERENCES

1. A Casale and R. S. Porter, Polymer Stress Reactions, Vols. 1 and 2, Academic Press, New York (1978).

2. R. P. Wool, J. Polym. Sci., Polym. Phys. Ed., 19: 449 (1981).

3. G. Bayer, W. Hoffman, and H. W. Siesler, Polymer, 21: 235 (1980).

4. R. L. Levy, Chem. Eng. News, 58(37): 51 (1980).

5. L. Penn and F. Milanovich, Polymer, 20: 31 (1979).

6. D. K. Roylance, Applications of Polymer Spectroscopy (E. G. Brame, Jr., ed.), Academic Press, New York (1978).

7. M. A. Grayson and C. J. Wolf, Applications of Polymer Spectroscopy (E. G. Brame, Jr., ed.), Academic Press, New York (1978).

8. G. D. Mendenhall, ANGEW. Chem., Int. Ed. Engl., 16: 225 (1977).

9. G. A. George, G. T. Egglestone, and S. Z. Riddell, Polym. Eng. Sci., 23: 412 (1983).

10. D. L. Fanter and R. L. Levy, ACS Sym. Ser., 95: 211 (1979).

11. D. L. Fanter and R. L. Levy, Chemtech, 9: 682 (1979).

12. J. I. Zink, Accts. Chem. Res., 11: 289 (1979).

13. A. N. Streletskii and P. Yu Butyagin, Polym. Sci. USSR, 15: 739 (1973).

14. P. Yu Butyagin, V. S. Yerofeyev, I. N. Musalyelyan, G. A. Patrikeyev, A. N. Streletskii, and A. D. Shulyak, Polym. Sci. USSR, 12: 330 (1970).

15. V. E. Krauya, Z. T. Upitis, R. B. Rikards, G. A. Teters, Ya. L. Yansons, Mekh. Kompoz. Mater., 2: 325 (1981).

16. G. A. George, G. T. Egglestone, and S. Z. Riddell, J. Appl. Polym. Sci., 27: 3999 (1982).

17. J. H. Richardson, S. B. Monaco, J. D. Breshears, D. C. Johnson, S. M. Lanning, and R. J. Morgan, "Stress Chemilumi- nescence: Predictive Applications to Polymer Failure," Proc. Division of Organic Coatings and Plastic Chemistry, Vol. 46, 183rd ACS National Meeting, Las Vegas, pp. 528–531 (1982).

18. S. B. Monaco, J. H. Richardson, J. D. Breshears, S. L. Lanning, J. E. Bowman, and C. M. Walkup, I&EC Prod. Res. Devel., 21: 546 (1982).

19. S. B. Monaco and J. H. Richardson, Polym. News, 9: 230 (1983).

20. D. L. Fanter and R. L. Levy, "Chemiluminescence of Stressed Polymers: Potential for Elucidation of Mechanochemical Contri- butions to Polymer Aging," Proc. Division of Organic Coatings and Plastic Chemistry, Vol. 39, 176th ACS National Meeting, Miami Beach, pp. 599–602 (1978).

21. R. L. Levy and D. L. Fanter, "Stress Chemiluminescence of Epoxy Resins," Proc. Division of Organic Coatings and Plastic Chemistry, Vol. 41, 178th ACS National Meeting, Washington, D.C. (1979).

22. C. H. Brennenstuhl and D. L. Fanter, "A Computer Controlled Chemiluminescence System," Proc. DECUS U.S., Miami Beach (1981).

23. C. J. Wolfe, D. L. Fanter, and M. A. Grayson, "Chemilumi- nescence of Thermosetting Resins," Proc. ACS Symposium on the Chemorheology of Thermosetting Polymers, Kansas City, MO (1982).

24. D. L. Fanter, Rev. Sci. Instrum., 49: 1005 (1978).

25. S. B. Monaco, L. D. Davis, and J. H. Richardson, J. Appl. Polym. Sci., 29: 4439–4442 (1984).

26. S. B. Monaco, J. H. Richardson, and L. D. Davis, J. Haz. Mat., 9: 31 (1984).

27. H. H. Kausch, Polymer Fracture, Springer-Verlag, Heidelberg, Chap. 6 (1978).

7
Radiothermoluminescence and Transitions in Polymers

LEV ZLATKEVICH   VIBO Research, Inc., Pennsauken, New Jersey

∠ωb p. 308

I. THERMOLUMINESCENCE IN POLYMERS INDUCED BY RADIATION

Luminescence observed as a result of the irradiation of a substance at a low temperature and its subsequent warmup is called radiothermoluminescence, and the resulting plot of emission intensity against temperature (or warming time) is called a glow curve.

In general, it seems that there can be three main causes of radiothermoluminescence in solid materials: excitations, chemical reactions, and charge trapping followed by recombination [1].

Molecular excitation has been found to be a very common cause of organic luminescence. The greatest objection to this as a cause of radiothermoluminescence is the comparitively short decay half-lives involved. The maximum half-lives observed in a wide variety of organic materials lasted a few seconds [2], whereas the half-life of radiation-induced luminescence at liquid nitrogen temperature is many orders of magnitude larger (samples usually give considerable emission even if kept for many days between irradiation and warming).

The most obvious chemical reaction mechanism would be radical-radical recombination. This can be ruled out as the cause of luminescence on a number of grounds [4]. The variation in thermoluminescence intensity with dose usually reaches a maximum at a few megarads, but radical production is linear with dose up to more than 50 Mrad [3]. No emission is observed when radicals are produced at low temperature by purely mechanical means (grinding) and the sample is then warmed up. The thermoluminescence output can usually be reduced to nearly zero by illuminating the sample with visible light prior to warming (optical bleaching), whereas this has no effect on radical concentration. The electron spin resonance investigation of alkyl radical concentration in relation to thermoluminescence emission showed that there is no increase in the rate of disappearance of alkyl radicals in the temperature neighborhood of the thermoluminescence peaks and that alkyl radicals continue to decay long after the luminescence is exhausted [4]. The kinetic of radical-radical recombination is expected to be of the second order, whereas radiothermoluminescence has been shown to follow first order kinetics. In addition, free radicals are generally completely "frozen in" at liquid nitrogen temperature, so they cannot react with each other. Moreover, the very small activation energy of the thermoluminescence process at temperatures not essentially higher than liquid nitrogen temperature makes any type of chemical reaction unlikely as the source of the thermoluminescence.

The only remaining possible cause of radiothermoluminescence in organic solids is the recombination of trapped ions. All phenomena accompanying radiothermoluminescence are explicable if trapped ions are the reactive species, and there is no doubt that thermoluminescence in irradiated organic substances is caused by ion recombination.

The term radiothermoluminescence, strictly speaking, presumes excitation by ionizing radiations and thus peculiarities of the interaction of high-energy radiation with matter are important in defining the radiothermoluminescence mechanism. The term high-energy radiation is applied both to particles moving with high velocity— fast electrons or β particles, fast protons, neutrons, α particles, and charged particles of higher mass—and to electromagnetic radiation of short wavelength—x-rays and γ-rays. The processes by which these different forms of radiation react with the atoms of a specimen through which they pass may be different, the common feature being the large amount of energy carried by each particle or photon (this energy is very much greater than that binding any orbital electron to an atomic nucleus or an atom to its neighbor). In this respect, they differ from slow or thermal neutrons and from ultraviolet light, in which the energy carried per particle or photon is usually smaller than the ionizing energy of an atom or molecule. The effects produced by γ-rays or x-rays may be best understood as due to discrete high-energy photons, which may therefore also be considered as particles in this context. In most cases, the changes produced by incident radiation depend mainly on the total energy absorbed and very little on the type of radiation or its energy.

Ionizing radiation as it passes through matter excites and ion-
izes the molecules surrounding its trajectory. The secondary electrons which are extracted from the molecules during ionization progressively transfer their kinetic energy to the medium until they become thermic. A thermic electron will be at a distance r_t from the
ion from which it has been issued (parent ion). Several cases may
be cited:

(1) If the distance r_t is less than a certain critical distance
r_c to which the coulomb attraction is exerted by the parent ion (that
is, $r_t < r_c$), then the electron will be drawn toward the parent ion,
which it will neutralize and form into a highly excited molecule.
Such highly excited molecules formed by neutralization eventually
become inactive, whether it be by decomposition, internal conversion, collision, or a combination of two or more of these factors.
The return to the parent ion is very rapid. The time which elapses
between the process of ionization and the process of neutralization
can be estimated at less than 10^{-10} sec. If the neutralization is
followed by an emission of light, this emission will occur 10^{-7} to
10^{-8} sec after ionization if the emission takes place beginning with
a singlet state and 10^{-3} sec to several minutes if it is from a triplet
state.

(2) If $r_t > r_c$, the electron breaks free and may recombine,
not necessarily with its parent ion but with any of the ions present
in the medium. Under normal conditions, the time between ionization and neutralization may be estimated at 10^{-2} to 10^{-3} sec. In
the two cases just examined, light emission is produced during irradiation. Such an emission is called radioluminescence.

(3) During and particularly at the end of the thermolization
process, the electron may become "trapped," and its recombination
with an ion (and, consequently, the emission of light) may be slowed
down. This retardation is highly variable. In the case of solid
media, it may be measured in minutes, hours, or sometimes even
days. It is this process, described as deferred radioluminescence,
with which we are presently concerned and which we will be examin-
ing in more detail.

Regardless of whether or not the secondary electron is temporar-
ily trapped, it will in time recombine with a positive ion and charge
neutrality will again be established. This event, as well as the reactions induced by the electron during its travel through the matrix,
will lead to a proliferation of stable and unstable products. Some
of the electronically excited molecules produced in this manner will
return to their ground state by luminescence emission and by inter-
or intramolecular energy transfer. The photons given off by the
sample can thus come from the excited molecule or from chemicals
attached to or contained in the polymer. In many cases, the polymer does not give any significant luminescence and the observed
emission is then from an impurity or additive molecules. Although

the actual luminescence is due largely to chemical impurities, the charge trapping is due mainly to the basic polymer structure. The concentration of impurity molecules is often extremely low, and thus most of the initial ionization and excitation will occur in the polymer itself followed by charge and/or excitation energy transfer to the luminescence centers, causing these to excite. The question then arises as to how the energy available after charge recombination is transferred to these centers. One possibility is that both electrons and holes recombine within the polymer matrix and the energy is then transferred by an excitation mechanism to the luminescence center [5]:

$$M^+ + L + e^- \longrightarrow M^* + L \longrightarrow M + L^* \longrightarrow M + L + h\nu \qquad (1)$$

In Equation (1), M and L are matrix and luminescence molecules, respectively, and the asterisk denotes an excited state. Alternatively, one of the charges migrates to and is trapped by the luminescence center, to be joined by the opposite charge when this is thermally detrapped:

$$M^+ + L + e^- \longrightarrow M^+ + L^- \longrightarrow M + L^* \longrightarrow M + L + h\nu \qquad (2)$$

A third possibility is that ionization occurring on the luminescence center results in the loss of a charge, which is trapped within the polymer matrix. This is then recaptured after thermal detrapping on warming:

$$L^+ + e^- + M \longrightarrow L^+ + M^- \longrightarrow M + L^* \longrightarrow M + L + h\nu \qquad (3)$$

II. RADIOTHERMOLUMINESCENCE OF ORGANIC SUBSTANCES AND MOLECULAR MOTION

The most immediate electronic processes of ionization, excitation, and electron capture are not influenced to any significant extent by the viscosity of the medium. Turning now to the processes which follow excitation and ionization, those involving molecular dissociation and reaction without mass transfer do not seem to be greatly influenced either by the rigidity of the medium or by molecular size [6].

However, all processes which depend on atomic (hydrogen atom) or molecular diffusion are profoundly affected by the viscosity of the matrix. After charge separation has been achieved, the subsequent reactions of stabilized electrons appear to be governed largely by molecular or segmental diffusion. The practical importance of

molecular motion is illustrated by the considerable variation in the
lifetimes of stabilized electrons according to the viscosity of the
medium.   Macromolecular systems are usually characterized by a
high viscosity, especially in the solid state at low temperatures.
Consequently, there is a large probability that intermediate species
generated during the irradiation of polymers remain trapped in the
host material, their reaction rates being essentially limited by the
slow diffusion process that applies in such circumstances.   In particular, stabilized electrons are generally produced in condensed systems by the polarization of surrounding molecules.

Traps in polymers can be of different natures, either physical
or chemical.   An electron physically trapped in a polymer matrix
can be considered as a distinct entity by itself, i.e., it is not localized in a specific molecular orbital belonging to a single molecule
or its segment.   Such an electron is localized in cavities or voids
which are associated with structural imperfections in disordered
solids and is only weakly coupled to its environment.   The concentration and distribution of physical traps depend to a great extent on
the sample prehistory.   One should expect a continuous depth distribution, because of various degrees of perfection for the same
type of a trap, as well as exponential (or pseudoexponential) distribution, because large structural defects are formed with an essentially lower probability than small ones.

Differently from a physically trapped electron, a chemically
trapped electron can be regarded as being bound to some particular
molecule and to reside entirely within the characteristic molecular
orbitals of that molecule.   A chemically trapped electron would therefore be expected to undergo significant interaction with the protons
(or other magnetic nuclei) of its host molecule.   Chemical traps are
characterized by a discrete depth distribution, reflecting the electron affinity of the various atomic groups representing chemical
traps.   A second peculiarity of traps of a chemical nature is the independence of their concentration and distribution of the conditions
of sample preparation (quenching, annealing, rate of temperature
change, etc.), although free radicals and some volatile impurities
(dissolved gases) are the exception.

Free radicals are the most important chemical traps.   The way
in which free radicals may influence electron trapping is by the
formation of a carbanion $R^{\cdot -}$:

$$R^{\cdot} + e^{-} \rightarrow R^{\cdot -} \tag{4}$$

It should be pointed out that reactions of the type represented by
Equation (4) can be expected to be exothermic by about 1 eV or
more [7], as judged by analogy with the electron affinity of the
methyl free radical, and should therefore be thermodynamically
feasible.   The experimental support for the existence of carbanions

in irradiated hydrocarbons has been obtained by Ekstrom et al. [8] , and it was suggested that the reaction of the radical ions R $\cdot^-$ and molecular ions M$^+$, rather than electron hole recombination, is responsible for the formation of excited molecules and following luminescence.

In the radiothermoluminescence experiment, the ion recombination rate constant increases with temperature (time), while the number of particles taking part in the reaction decreases. Thus the number of photons emitted per second reaches a maximum due to these two competing effects. In a real system, ion recombination is usually not restricted to a single process, but there is a set of more or less sharply divided processes, which are characterized by their own recombinational parameters. As a result, the glow curve may exhibit several maxima.

Table 1 presents the positions of the thermoluminescence peaks for a certain number of organic compounds which are known to possess points (or regions) on the temperature scale at which either phase or relaxation changes occur [11—13]. It can be seen that for all these compounds, most thermoluminescence peaks correspond to the transition temperatures. This highly important observation was first made by Nikolskii and Buben [11] and later was generalized by Semenov [13], who related it to an analogous observation concerning the recombination of radicals. There is, nevertheless, some ambiguity in the results: a certain number of peaks appear at temperatures other than those of the known transitions. The question that must be asked is whether these peaks correspond to any occurrences whatever in the lattice.

In order to answer this question, one may consider the effect produced by the addition of substances possessing an electron affinity, such as biphenyl and naphthalene, to cyclohexane and ethanol. It has been noted that such an addition increases the intensity of radiothermoluminescence and modifies its spectrum but does not change the position of the peaks [12].

The only kind of trap which exists in saturated hydrocarbons is a "hole" formed as the result of electron polarization of the surrounding medium with its induced moments. The energy (or depth) of the trap may be estimated to several tenths of an electronvolt and may be shown to be at maximum if electrons happen to be in a "defect" (cavity) in the lattice. If the substance is polar, the trap may be formed by an abnormal orientation of the dipoles, pointing to a cavity. Such defects may already exist in a rigid medium or may be created by the electron itself if the type of medium permits the reorientation of dipoles. The depth of the trap may then reach 1 to 2 eV. Finally, for the medium made up of (or containing in dissolved form) molecules having electron affinity or radicals (these compounds being present initially or, as is the case with radicals, created by irradiation), these entities may intercept the electrons

with the formation of ions or of negative ion radicals. In this
event, the depth of the trap is determined by the electron affinity
of the compound in question (a range of 0.4 to 0.7 eV for polyaromatic hydrocarbons and 1 to 2 eV for radicals).

Thus it can be concluded that the depth of the traps is different in pure cyclohexane and ethanol, where it is mainly a question
of solvated electrons, and in the presence of polyaromatic compounds,
where negative ions may be formed. Therefore, electron release
from the trap does not consist of electrons evaporating from their
traps but must rather consist of a mechanism able, on the one hand,
to liberate solvated electrons (ones trapped in cavities) and, on the
other hand, to bring about the rapid diffusion of molecular ions
having dimensions on the same order as those of the molecules com-
posing the lattice. We are thus led to believe that all thermoluminescence peaks coincide with any and all reorganizations of the molecular network. The other evidence supporting such a conclusion is
provided by the detailed study of thermoluminescence in cyclohexane
having undergone varying thermal treatments and the correlation of
the temperature of the peaks with crystallographic, calorimetric,
and other kinds of data [12]. This study allowed the linking of at
least three of the four thermoluminescence peaks from this substance
with reorganization of the crystal lattice.

Along with the low molecular weight substances, the analogy between the positions of thermoluminescence peaks and structural
transitions can be readily appreciated in polymers as well. A comprehensive analysis of dilatometric, calorimetric, dynamic-mechanical,
and electron spin resonance data led Boyer to the conclusion that
polyethylene exhibits three low-temperature transitions at about
150, 190, and 240L [14]. As a rule, none of the methods taken
separately is sufficient to resolve all three transactions. By means
of radiothermoluminescence, however, three transitions whose positions coincide well with Boyer's estimate are observed [15].

It was noted that factors affecting the position of the glass
transition and other transitions in a polymer (cross linking, plasticization, preorientation of the polymer film, heating rate, etc.) also
affect in a similar way the corresponding peaks on the radiothermoluminescence curve. Moreover, the activation energies for radiothermoluminescence and molecular motion in polymers are in good
agreement over the whole range of temperatures studied [16—22].

All this indicates that molecular motion is a determining factor
in the destruction of active products of radiolysis in organic substances; the rapid release and recombination of stabilized active
particles is possible only by thawing out the molecular mobility in
the regions of temperature transitions. The similarity of the activation energies for radiothermoluminescence and molecular mobility
over the same temperature range serves as the basis for the hypo-
thesis that the lifetime of charges stabilized on structural elements

TABLE 1 Transitions in Organic Substances

Substance	Phase	Transition temperatures (K)	Radiothermoluminescence peaks (K)	Nature of the transition
Methyl alcohol	cr	157	120, 155	
Ethyl alcohol	am	110	103–107	
	cr	110^+, 156^*	113, 143	$^+$Recrystallization *Melting
n-Butyl alcohol	cr	160^+	110, 152	$^+$Phase transition of first order
	am	120–130^+	123	$^+$Glass transition
Benzene	cr	110^+	129, 190, 223	$^+$Onset of rotation of the molecule
Hexamethylbenzene	cr	108^+, 135–165^*	117, 144, 180, 220	$^+$Phase transition *Onset of rotation of the molecule
Cyclohexane	cr	133–183^+, 186^* 218^-	114, 158, 186, 202	$^+$Onset of rotation of the molecule *Crystal latice rearrangement $^-$Onset of self-diffusion

1,1-Dicyclohexyl-dodecane	cr	300^+	137, 302	$^+$Melting
	am	190^+	140, 197	$^+$Glass transition
n-Octadecane	cr	297^+, 301^*	162, 293	$^+$Phase transition of first order *Melting
Paraffin	cr + am	152, $230-250^+$, 333^*	152, 223, 330	$^+$Glass transition *Melting
Polyethylene	cr + am	$173-208^+$, 236^*	$123-165$, $175-208$, $237-248$	$^+$Onset of rotation *Onset of segmental motion
Polyisobutylene	am	221^+	155, 228	$^+$Glass transition
Teflon	cr + am	168, 295^+	150, 295	$^+$Phase transition of first order

Source: Refs. 11 to 13.

(or close to them) is in general proportional to the relaxation times of structural elements in question. The specificity of molecular motion, namely, its relaxational nature, governs the kinetics of recombination of charges in amorphous and semicrystalline organic substances as well as the radiothermoluminescence process. The relaxational character of molecular mobility leads to a reorientation of certain structural elements even at temperatures below those of temperature transitions registered by radiothermoluminescence. As a result, an isothermal luminescence occurs after irradiation. An increase in temperature leads to a decrease in the relaxation time of structural elements, to a decrease in the lifetime of charges stabilized on these structural elements, and to an increase in luminescence intensity. Other conditions being equal, the larger the number of structural elements of a given type in a polymer, the more charges will be released on thawing out of the mobility of these structural elements and the higher will be the intensity of the corresponding radiothermoluminescence peak. One may expect that the radiothermoluminescence maxima will be located in the same temperature intervals as transitions observed by the other methods of analysis.

With amorphous polymers, some of the stabilized charges have already recombined at low temperatures as a result of the beginning of motion of side groups or small segments of macromolecules in the irradiated sample. The majority of ions recombine, however, near the glass transition point, thus resulting in the appearance of the most intensive radiothermoluminescence maximum. Figure 1 shows the glow curve of a random styrene-butadiene copolymer containing 30 wt% styrene (curve e). Here also the results of evaluation are given of molecular relaxation in this copolymer by several other methods. The glass transition temperature of the copolymer was found to be 211K, as determined from the position of the intensive radiothermoluminescence maximum in the glow curve. This value differs from that measured by other methods by less than 5°.

Similar results were obtained for a great number of amorphous and semicrystalline polymers; the temperature corresponding to the most intensive radiothermoluminescence maximum practically coincides with the glass transition values determined by relaxational spectroscopy at frequencies of 0.01 to 0.1 Hz [23]. Besides, radiothermoluminescence provides one of the most sensitive techniques for evaluating secondary temperature transitions located below the glass transition and in some cases it also can be used for the analysis of the transitions above the glass transition.

This rather interesting peculiarity of the radiothermoluminescence process, namely, the occurrence of luminescence flashes at temperatures corresponding to those of temperature transitions, has finally led to the creation of a new method of analysis or organic solids and polymers—the radiothermoluminescence method.

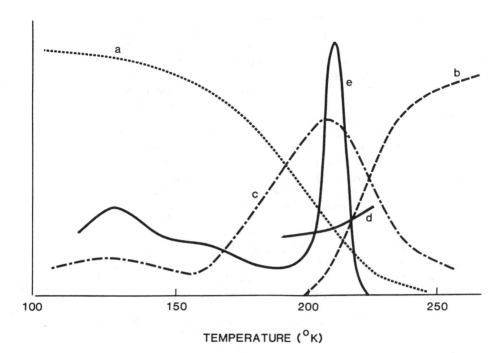

TEMPERATURE (°K)

FIGURE 1   Evaluation of the relaxation transitions in a styrene-
butadiene random copolymer (30% styrene) by means of various
methods: (a) nuclear magnetic resonance absorption line width;
(b) tensile compliance; (c) mechanical losses at 0.1 Hz; (d) dilatometry; and (e) radiothermoluminescence glow curve.   (After Ref.
20, with permission of the Rubber Division of the American Chemical Society.)

III.   INFLUENCE OF IMPURITIES ON
RADIOTHERMOLUMINESCENCE

The question of the influence of impurities on the radiothermoluminescence process in polymers is of major importance from both theoretical and practical points of view.   One of the most significant
conclusions agreed on by many researchers is the relative independence of the temperature of the glow peaks and overall shape of the
glow curve from the nature and concentration of additives.

    Charlesby and Partridge [24] reported that the impurities pres-
ent in polyethylene have no effect at all on the thermoluminescence
emission, since the glow curves of low-density polyethylene specially
manufactured without any additives and commercial materials which
contain 0.5% antioxidants were virtually identical.

Fleming [25] found that poly(methyl methacrylate) containing a considerable range of polymerization initiators and chain transfer agents exhibited a glow curve similar to the one for carefully purified methyl methacrylate monomer liquid polymerized by γ radiation. He also showed that the form of the glow curve is unaltered when polystyrene is doped with 1% by weight of benzophenone, 1,4-cyclohexane dione, or benzophenone oxime (anthracene) [26].

Applying the radiothermoluminescence technique to the study of 1,2-polybutadiene, Bohm [27] found that the luminescence intensity increases with greater purity of the material whereas the shape of the curve remains unchanged. This suggests that the luminescence species are incorporated into the polymer during the polymerization process or by subsequently occurring grafting reactions in contrast to the apparently excitation-quenching impurities which can be removed by the purification.

A comparison of the thermoluminescence curves for a variety of elastomers and their blends with and without the ingredients of vulcanization led Buben et al. [28] to the conclusion that the addition of ingredients has no effect on the shape of the light emission curves, although it sharply reduces the intensity of luminescence (by the order of 3 to 4).

A comprehensive study of the influence of three different dyes [crystal violet (CV), phenanthrene quinone (Ph), and rhodamine 6G (Rh)] on the glow curves of a polybutadiene (Solprene 233) and a styrene-butadiene block copolymer (Solprene 416) was carried out by Zlatkevich et al. [29]. Doping was performed by two methods: (1) the polymer was kept in the solution (methanol/toluene = 9:1) saturated with the dye for 28 hr and then dried in air, and (2) 0.1 wt% of the dye was roll-milled directly into the polymer.

Similar glow curves have been obtained for the samples doped by swelling and roll milling for all polymer—dopant systems studied. Practically no differences in the overall glow curve shape have been found between the glow curves for undoped samples and those for samples doped with Rh and CV. Ph-doped samples showed a noticeable relative decrease in glow intensity in the temperature region from 120 to 180K (Figure 2). However, it is important to note that the doping with Ph was accompanied by a change in sample color during and just after preparation. For instance, Solprene 416 roll-milled with Ph turned from yellow (the color of Ph) to pink. Since the polymers with physically dispersed CV and Rh showed glow curves similar to those of undoped samples, the differences in relative thermoluminescence intensity between undoped samples and samples doped with Ph were attributed to the occurrence of a chemical reaction between Ph and other additives contained in or attached to the polymer. It has to be underlined that the position of the main thermoluminescence maximum (T_g region) was found to be independent of either the polymer-dye combination or the method of incorporation of the dye (Table 2).

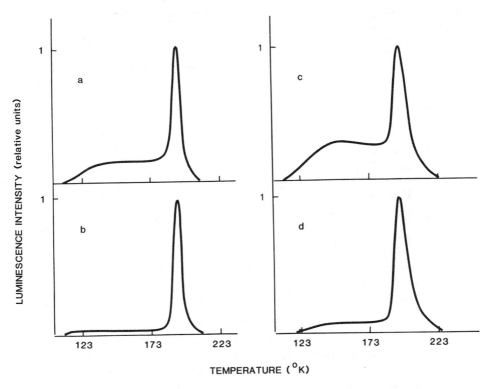

LUMINESCENCE INTENSITY (relative units)

TEMPERATURE (°K)

FIGURE 2 Glow curves for (a) Solprene 233, (b) Solprene 233
doped with Ph, (c) Solprene 416, and (d) Solprene 416 doped with
Ph. (After Ref. 29, John Wiley and Sons, with permission.)

The picture, however, will not be complete if several contradic-
tory reports are not mentioned here. Using UV light as the excitation source, Linkens and Vanderschueren came to the conclusion
that the type of impurity contained in a polymer can play a prevailing role in thermoluminescence and that an observed glow curve is
often typical of the dopant—polymer system [30,31]. However,
their experiment cannot be considered as a pure radiothermoluminescence test, since the latter requires high-energy ionizing irradiation. Furthermore, differences should be expected between the
thermoluminescence curves after exposure to ionizing and UV irradiation, especially for polymer—dopant systems. The first interacts
with the whole matrix in a statistically equivalent manner (no selectivity of absorption peculiar to one particular type of molecule),
whereas the latter is absorbed only by individual chromophores and
aromatic impurities.

TABLE 2   Temperatures of Thermoluminescence Maximum for Undoped and Doped Polymers

Samples	Temp. (K)	
	Roll milling[a]	Solvent immersion[a]
Solprene 233	194	193.5
Solprene 223 and CV	194	193
Solprene 233 and Rh	193.5	194
Solprene 233 and Ph	194.5	194
Solprene 416	195	195
Solprene 416 and CV	195.5	195
Solprene 416 and Rh	195	195.5
Solprene 416 and Ph	195.5	195

[a]Method of preparation.
Source: Ref. 29.

Results indicating that the glow curve of polyethylene with an
additive does not coincide with the glow curve of the pure polymer
were published by Lednev et al. [32].   In this study, different additives (phenanthrene, naphthalene, and anthracene) were introduced into polyethylene from the saturated vapor at room temperature, and it was reported that the positions of the individual maxima
and the shape of the glow curve are affected not only by the type
of the additive but by variations in its vapor pressure as well.

First of all, it has to be underscored that semicrystalline polymers (especially polyethylene, with its broad variety of different
structural formations) are probably not appropriate subjects for
studying the influence of additives on radiothermoluminescence glow
curve because any changes in the glow curve may result from the
structural changes connected with additive introduction.   Even when
a solvent as an intermediate for the introduction of an additive is
omitted and the additive is introduced from the vapor phase, such
a procedure cannot be visualized as simply the filling of the free
volume of a polymer by additive molecules.   Structural rearrangements are expected to take place and can be thought to depend on
many factors: the degree of equilibrium of structural formations
prior to additive introduction, polymer-additive affinity, the molecular size and concentration of additive molecules, the nucleating efficiency or plasticization ability of the additive, etc.   However,
there are no obvious reasons for the additive, which presumably
creates changes in the thermoluminescence curve of a semicrystalline

polymer, not to do so when introduced into an amorphous material of a similar chemical nature. The additive which serves as a charge trap in a hydrocarbon semicrystalline polymer (only in this case a variation in the thermoluminescence curve not related to structural changes can be expected) should play the same role in an amorphous hydrocarbon polymer. It has to be noted, however, that the stabilization of one of the charges on additives is the necessary, but not sufficient, condition for any change in the glow curve. Meggitt et al. [33] showed that in polyethylene doped with biphenyl the most probable radiothermoluminescence mechanism involves electron capture by the biphenyl, but luminescence occurs when the cations trapped in the polymer matrix are released and migrate toward the biphenyl anions. The recombination of charges in this case is directly related to "unfreezing" of the molecular mobility in the matrix.

To summarize, one may conclude that at the present time there is no clear evidence proving the prevailing role of additives on the shape of the glow curve, provided that introduction of the additive was not accompanied by structural changes in the matrix. On the contrary, the results obtained with a variety of polymers show quite convincingly that additives dispersed into a polymer matrix do not have any significant influence on the position of the thermoluminescence maxima and the overall glow curve shape. Consequently, radiothermoluminescence can be applied as a method for transition analysis in polymers without interferences owing to possible sample contamination.

IV. EVALUATION OF THE ACTIVATION ENERGY OF MOLECULAR MOTION BY RADIOTHERMO-LUMINESCENCE

Many methods have been proposed for obtaining activation energies from glow curve peak temperature and shape. Most of these assume either first or second order kinetics, and most also demand a glow peak which is well separated from its neighboring peaks. Furthermore, it is assumed that neither activation energy nor luminescence constant (the probability of photon emission from each recombination) varies during the extent of the peak. Since the methods for evaluating activation energy originally were developed for inorganic substances, it remains to establish the meaning in the polymer context, to the terms trap depth and frequency factor found in discussions of thermoluminescence in inorganic materials. For inorganic materials, trap depth is often taken to be the electron affinity of the electron trap, whereas the frequency constant has been visualized as the attempt frequency of the electron trying to escape from its trap (the trap is tacitly assumed to be immobile).

Developing the concept that electrons trapped in polymers at low temperature are released not by thermal activation but by the "unfreezing" of molecular motion, Partridge [34], instead of considering the electron attempting to gain sufficient energy to escape from an immobile trap, pictured the electron as somewhat loosely bound to a segment of a vibrating (rotating) molecular chain which is attempting to "shake it off" when it jumps between equilibrium positions. Such jumping motions are commonly referred to as primary and secondary relaxations in the context of dielectric and dynamic-mechanical studies.

The relationship between the intensity of thermoluminescence, activation energy, and temperature can then be written as

$$I = \alpha P n \, \nu o \, \exp\left(\frac{-E}{kT}\right) \qquad (5)$$

where
    P = probability per vibration for a trapped electron to be released
    νo = jump frequency of the molecular chains
    α = luminescence constant
    n = number of trapped electrons
    On the other hand, the intensity of the first order recombinational thermoluminescence in inorganic materials can be expressed as

$$I = \alpha S n \, \exp\left(\frac{-E}{kT}\right) \qquad (6)$$

Equation (5) is identical to (6), except that the frequency factor S in (6) has been replaced by $P\nu o$ in (5). According to (5), the differences in the glow peak positions are due to the different degrees of molecular motion in different structural regions and the activation energies of the thermoluminescence peaks are in fact the molecular motion activation energies. The fact that Equations (5) and (6) have precisely the same form indicates that the methods of glow curve analysis commonly applied to inorganic materials can be confidently used for organic substances.

A. Methods Employing Shape Parameters of the Peak

The most widely used methods are summarized in Table 3. A number of other methods should be briefly mentioned. Land [35] suggested a method which uses, in addition to T_p, the two inflection points in the thermoluminescence curve rather than the half-intensity temperatures T_1 and T_2. The method based on measurements of T_1, T_p, and T_2 is that of Keating [36]. Maxia et al. [37] developed a method for evaluating the activation energy and frequency factor for a

TABLE 3 Various Methods of Calculating Activation Energy

Study	First order kinetics	Second order kinetics	Limitations
Grossweiner [42]	$E = 1.51kT_p T_1 (T_p - T_1)$		$E/kT_p > 20$ $3T_1 \exp(E/kT_p)/[2T_p (T_p - T_1)] > 10^7$
Luschik [43]	$E = kT_p^2 (T_2 - T_p)$	$E = 2kT_p^2 (T_2 - T_p)$	
Kelly [44]	$E = 1.461kT_p T_1/(T_p - T_1)$	$E = 1.763kT_p T_1$	$E/kT_p \gg 1$ $E/kT_p \gg 1$
Halperin [45]	$E = [q/(T_2 - T_p)]kT_p^2$	$E = [q/(T_2 - T_p)]kT_p^2$	$E > 10kT_p,\ q < 1$ $1 \leq q \leq 2$

Note: q, peak symmetry factor; T_p, peak temperature; T_1, temperature on the low-temperature side of the peak at which the luminescence intensity attains it half-maximum value (K); T_2, temperature on the high-temperature side of the peak at which the luminescence intensity reduces to its half-maximum value (K); k, Boltzmann's constant (eV/K).
Source: Ref. 25.

multiple-peak glow curve in which the various peaks result from the release of electrons from a single trap and their recombination with various recombination centers. Onnis and Rucci [38] discussed the alternative possibility of obtaining several glow peaks, namely, having several traps and a single recombination center.

It should be noted that all the equations for calculating activation energies assume that the luminescence efficiency of the luminescence centers is independent of temperature. Such an assumption seems reasonable at low temperatures, where radiationless collisional deactivation of the excited states is essentially prevented. However, at relatively high temperatures, phosphorescence quenching may be of significance.

B. The Initial-Rise Method

Garlick and Gibson [39] suggested a method for activation energy evaluation known as the initial-rise method, which is usually considered to be more general than other methods because it is independent of kinetic order. If one studies Equation (6), one can see that at the beginning of the glow peak, n changes only slightly with temperature, and therefore $I \propto \exp(-E/kT)$. Thus plotting ln I as a function of $1/T$ should yield a straight line in this region, the slope of which is $-E/k$. The same reasoning can be applied to the thermoluminescence governed by the second order recombination. This method has further been developed by Gobrecht and Hoffman [41], who used subsequent heating and cooling cycles to obtain the "spectroscopy of the traps."

It should be noted that in the case when phosphorescence quenching is essential, the initial-rise method should yield an underestimated (reduced) activation energy value [40].

C. Various Heating Rates

Another group of important methods is that of various heating rates. For the first order kinetics, the temperature of the glow peak maximum can be obtained by differentiation of Equation (6), which for a linear warming rate of β gives

$$SkT_p^2/E \beta = \exp\left(\frac{E}{kT_p}\right) \tag{7}$$

Upon further differentiation of Equation (7) with respect to heating rate, one obtains

$$\frac{dT_p}{d\beta} = \frac{T_p^2}{2T_p + (E/k)} \tag{8}$$

From Equation (8) it is clear that an increase in heating rate will always shift the peak maximum toward higher temperatures and vice versa. Furthermore, this shift will be largest for glow peaks with the lowest activation energy. Bohun [42] and Parfianovitch [43] suggested that a sample should be heated at two different linear heating rates, β_1, and β_2, and that the corresponding peak temperatures T_{p1} and T_{p2} should be registered. Equation (7) can then be written once for β_1 and T_{p1} and once for β_2 and T_{p2}. If one divides these equations one by the other, one gets an explicit equation for the calculation of E:

$$E = [kT_{p1}T_{p2}/(T_{p1} - T_{p2})]\ln[(\beta_1/\beta_2)(T_{p2}/T_{p1})^2] \tag{9}$$

Hoogenstraaten [44] suggested the use of several (linear) heating rates; plotting $\ln(T_p^2/\beta)$ vs. $1/T_p$ should yield, according to Equation (7), a straight line from whose slope E/k, E is found.

It is to be noted that even in the case where temperature quenching is essential, utilization of one of the various heating rate methods would yield correct (not underestimated) activation energy value [40].

D. Isothermal Decay

The method of isothermal decay enables the measurement of E and S in the first order case. If one holds a sample at a constant temperature in a range where thermoluminescence appears during heating, one can measure the isothermal decay, which is given by the solution of Equation (6) for the T = constant case as follows:

$$I(t) = nS \exp\left(\frac{-E}{kT}\right) \exp\left[-St \exp\left(\frac{-E}{kT}\right)\right] \tag{10}$$

Plotting $\ln[I(t)]$ as a function of t would give a straight line (the occurrence of a straight line ensures the first order property) the slope of which is

$$M = S \exp\left(\frac{-E}{kT}\right) \tag{11}$$

Repeating the measurements at various temperatures, one gets various values of M. Plotting ln M as a function of $1/T$ should give a straight line with slope $-E/k$, thus enabling the evaluation of E and, by substituting into Equation (11), determination of the value of S. This method yields the value of E in the case of exponentially temperature-dependent recombination probability [45].

V. RADIOTHERMOLUMINESCENCE IN COMPARISON TO THE OTHER METHODS OF TRANSITION ANALYSIS

Radiothermoluminescence possesses numerous advantages over other methods for observing transitions in polymers. Among the advantages are the following:

1. Speed and relative simplicity of the analysis. With normal heating rates ($10°C/min$), a run is completed in a few tens of minutes.
2. The thermoluminescence of many irradiated organic substances is so intense that it can be observed with the naked eye. With photomultiplier tubes, it is possible to study the thermoluminescence of specimens whose weight does not exceed a few tenths of a milligram.
3. Samples of arbitrary configuration can be investigated, e.g., films, pellets, fibers, individual crystals, sliced sections, etc. Because sample shape/size requirements are minimal, strained specimens can easily be investigated. One simple device uses a metal ring to secure the stretched sample on a metal backing plate.
4. High accuracy and resolution of measurements. The radiothermoluminescence method gives a continuous curve of luminescence intensity vs. temperature, whereas many other methods (mechanical and dielectric spectroscopies, for instance) require repetitive measurements of a certain parameter at different temperatures and then plotting of this parameter vs. temperature. Determination of transitions by means of radiothermoluminescence is differential (recording of the peak positions), whereas nuclear magnetic resonance, volume dilatometry, and thermomechanical analysis give an integral curve (the transition corresponds to the inflection point). The temperature of the transition, as given by the peak position T_p, is obtained directly by the radiothermoluminescence experiment. Because of the narrowness of the peak, T_p usually can be determined with an uncertainty of less than $1°C$. This means that subtle effects resulting in shifts of the transition temperature by only $1°C$ can be studied by the thermoluminescence method.
5. It is well known that multiple transitions are best resolved by a low-frequency test method; transition peaks tend to merge with one another as the measuring frequency is increased. Although in static, nonisothermal experiments various molecular motion frequencies usually correspond to transitions located at different temperatures, the range of effective frequencies covered by radiothermoluminescence is somewhere between 10^{-4} and 10^{-1} Hz, thus permitting resolution of overlapping processes.

6. The half-width of the thermoluminescence peak (ΔT_p) also provides valuable information. If the half-width of the peak appreciably exceeds the half-width of the elementary single relaxation time maximum, ΔT_0, the difference $\Delta T_p - \Delta T_0$ characterizes the nonuniformity of the transition temperature throughout the volume of the specimen.

7. The activation energy of molecular relaxation can be measured over a broad temperature range.

Radiothermoluminescence is especially suitable for evaluating temperature transitions at low temperatures, although there are several reports stating the usefulness of the technique at temperatures essentially above room temperature [46,47].

An inherent disadvantage of the radiothermoluminescence method in its conventional setup is the need for a high-power source of irradiation. However, instruments with a self-containing irradiation source have recently been introduced [48,49].

VI. INSTRUMENTATION

The radiothermoluminescence method is noted for the relative simplicity of the installation for registering glow. One of the possible constructions is schematically illustrated in Figure 3. It consists of two parts [15]:

1. The cryostat
2. The optical system

The cryostat can be cooled down to $-196°C$ by liquid nitrogen circulation and heated up to $100°C$ with the heater.

The optical part of the installation consists of a photomultiplier tube and a diaphragm with a shutter. The diaphragm shutter is used to establish the "zero light level" of the photometer and to avoid its overloading.

The cryostat is separated from the optical part of the installation by the glass window. To avoid moisture condensation on the window, dry nitrogen is purged over it during the experiment. The spectral composition of thermoluminescence emission can be studied by replacing the glass window with a light filter. After the cryostat is cooled down to $-196°C$, the cuvette with a preirradiated sample is placed on the top of a sample holder and is fed into the cryostat, where it is in tight contact with the light guide. The circulation of liquid nitrogen through the cryostat is then shut down, the heater turned on, and the sample warmed up at a constant rate. The temperature of the sample is registered by a constantan-copper thermocouple and the intensity of emitted light is recorded continuously.

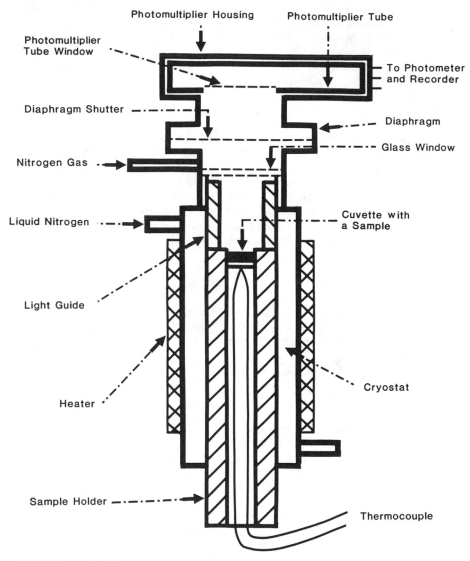

FIGURE 3 Radiothermoluminescence detector and cryostat. (After
Ref. 15, John Wiley and Sons, with permission.)

Irradiation of the sample prior to the analysis is usually carried
out with γ-rays from a ^{60}Co source, although x-ray and β-radiation
sources have been used. The dose rate is not an important factor.
All that matters is that the total dose be from 0.1 to several mega-

rads. As a rule, irradiation of a substance at a dose of 1 Mrad
causes no marked changes in its structure; hence the results can be
related to the initial nonirradiated substance.

It was recently shown that the radiothermoluminescence experiment can be performed using relatively low-energy electrons for irradiation [48]. A beam of electrons having an energy of 10 to 50
keV can be incorporated into the radiothermoluminescence installation
and does not need any special insulating precautions. An apparatus
which combines the irradiation source, cryostat, and optical system
is a simple construction. The need to have a vacuum chamber for
specimen irradiation and subsequent thermoluminescence analysis is
the only increase in complexity required.

The other, even more attractive opportunity was provided by
the development of a cryogenic scanning electron microscope (SEM)
stage which permits electron irradiation of samples and subsequent
thermoluminescence analysis (Figure 4). Irradiation can be performed with either fixed, defocused electron beam or a scanning
beam. The latter permits irradiation of only part of the total specimen and additionally gives more uniform irradiation over the sample.
The detection system utilizes a vacuum feedthrough manipulator for

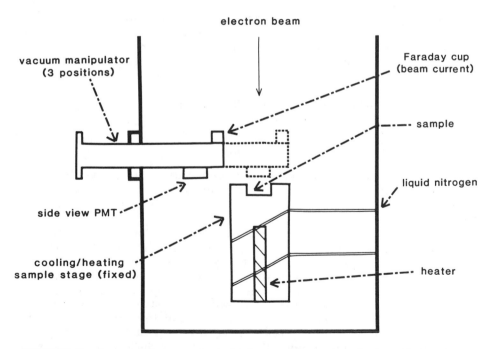

FIGURE 4 Cryogenic scanning electron microscope stage and vacuum
feedthrough manipulator.

positioning the photomultiplier tube before and after irradiation.
Other features of the system include very short irradiation times
(the dose of 1 Mrad can be achieved in seconds) and improved environmental control during heating (N_2, O_2, etc.). Also, varying
the electron energy permits control of penetration distance, thus
achieving a depth profiling not available with other spectroscopic
techniques.

In the process of developing RTL with electron-irradiated samples it was noticed that the relative intensity of high-temperature
RTL peaks is invariably larger than in the case of γ-irradiated samples analyzed by "conventional" RTL [49]. Investigation of this
seeming inconsistency led to the conclusion that it can be attributed
largely to O_2 quenching of excited states. (While O_2 is present
during conventional RTL analysis, the RTL/SEM experiment is conducted entirely in vacuum ($P < 10^{-4}$ Torr), so O_2 quenching is not
a problem.) This was confirmed when several materials (polystyrene, poly(methyl methacrylate), and nylon 6) were studied in a
broad temperature interval (from 77 to 473K). Although no peaks
in these polymers were registered above 273K by means of conventional RTL, numerous high-temperature peaks were observed with
the SEM.

Commercialization of the scanning electron microscope attachment
containing the cryogenic stage and the vacuum feedthrough manipulator promises to make the radiothermoluminescence method more convenient and widespread, and also broaden the range of polymers
which can be studied.

VII. APPLICATION OF RADIOTHERMOLUMINES-
CENCE TO THE STUDY OF POLYMER
SYSTEMS

A. Polybutadienes

It is well known that the chain microstructure of a polymer influences
its relaxation behavior. A case in point is the diene polymers,
which can contain essentially different concentrations of cis, trans,
and vinyl structures.

The glow curves of three polybutadienes having various vinyl
contents are shown in Figure 5. For all the materials, there are two
temperature ranges in which the luminescence intensity increases
rapidly. The first is 130 to 160K. In this temperature range, the
glow curves have complicated shapes. For samples A and B, two
adjacent maxima are superimposed, whereas sample C exhibits a
single, poorly resolved maximum. The second temperature range in
which there is a sharp increase in the intensity of luminescence is
160 to 273K. In this range, there is a clearly defined single maximum
whose position shifts to higher temperatures when the vinyl concentration increases [51].

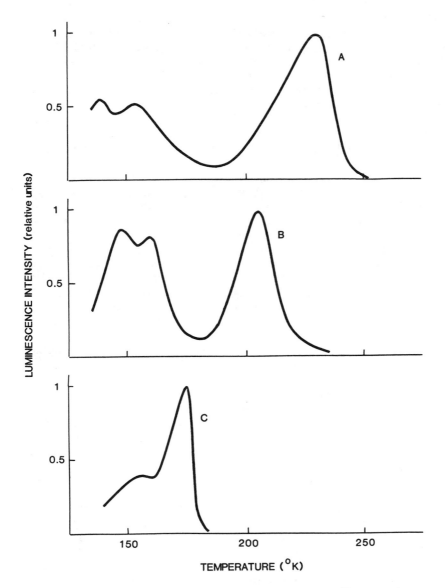

FIGURE 5 Glow curves of polybutadienes with different vinyl concentrations: (A) 66%; (B) 40%; and (C) 5%. (After Ref. 40,
Pergamon Journals, Inc., with permission.)

The presence of the maxima in light emission during warming
to room temperature is an indication that an accelerated liberation of
electrons from traps takes place at the temperatures at which these
peaks appear. Increases in the rate of detrapping can occur because the thermal energy kT is sufficient to lift an electron over
the barrier potential of the trap or because the trap itself is eroded
by the onset of structural changes occurring at this temperature.
As was indicated earlier, the latter process predominates in polymer
systems.

At least three different types of traps can be proposed for polybutadiene [52]. Consistent with the free-volume concept, it can be
assumed that fluctuations in density exist in the frozen matrix, mani-
festing themselves as cavities. Free electrons trapped in some of
these abundant low-lying energy wells should easily escape as a result of the supply of only modest amounts of energy. The removal
of electrons from deeper cavity traps would require more energy;
however, this could be accomplished even at low temperatures by
the onset of vinyl group rotation about C—C bond which links this
group to the main chain or by a rotation of short-chain segments in
"crankshaft" configurations. The latter should be possible well below
T_g and should have a low activation energy. Also contributing can
be the vibration of chain segments or submolecules about their equilibrium positions.

A second possibility is for the electron to be attached to double
bonds having a positive electron affinity. Since unsaturation and,
in particular, vinyl groups are present in polybutadienes, they too
represent available electron traps.

Finally, transient intermediates as radicals and radical ions must
be considered as traps. Since the ionization potential of π electrons
is considerably lower than that of σ electrons in the aliphatic chain,
it can be assumed that the radical ions formed by irradiation are
predominantly located in the vinyl groups [52].

An unequivocal correlation of the peaks shown in Figure 5 with
the types of traps present in polybutadiene is not possible at this
time. However, the low-temperature transitions in the 130 to 160K
region (secondary relaxations) most probably should be assigned to
cavity traps which are destroyed by motions involving pendant vinyl
groups and short segments of the backbone structure. As Figure 5
illustrates, the intensity of secondary relaxation is larger for polymers with high vinyl concentrations. This is predictable, since a
high vinyl content inevitably results in a structure with a larger
concentration of defects.

Besides secondary relaxation around 160K, a fairly broad peak
at 118K has been reported for high vinyl content polybutadiene [52].
In this study, irradiation was carried out at 90K, and thus the difference between the irradiation temperature and the position of the
first low-temperature maximum was only 28K. It must be born in

mind that at the beginning of the heating period the intensity of radiothermoluminescence increases independently of the presence of a transition and that the initial rise in emission is due apparently to the increase in the rate of interaction of the ionized states on heating in comparison with their rate of interaction at the temperature of irradiation. Thus in some cases the first low-temperature maximum (around 90 to 120K) is only apparent; its position and shape are determined by the temperature and regime of irradiation. One of the ways to discriminate between a true relaxation peak and an apparent peak not directly related to relaxation is to perform irradiation and the start of subsequent heating at still lower temperatures (e.g., at liquid helium temperature instead of liquid nitrogen temperature).

Along with secondary relaxation peaks, each of the polybutadienes exhibits the most intensive maximum at temperatures approaching the glass transition. It is believed [52] that all the ion recombinations which could take place through detrapping of the electron and subsequent migration toward the positive charge have occurred. Remaining in the sample are trapped electrons which cannot be liberated by supplying energy of the amount of kT or by the erosion of traps induced by the onset of local motion. While it is possible that certain cavity traps remain intact up to T_g, it is more likely that these deep traps are associated with free radicals, which are known to possess a high electron affinity. Charge recombination involving these groups then occurs at T_g, not by detachment of the electrons but by a physical approach of the two molecular ions.

Polybutadiene samples with greater vinyl concentrations exhibit higher T_g values. Moreover, there is a direct proportionality between T_g values and vinyl concentration (Figure 6).

The T_g position for all polybutadienes depends on the heating rate, shifting to higher temperatures when heating rate is increased. Partridge [1] applied Equation (7) to the results obtained by Alfimov and Nikolskii [51] on the variations of glass transition glow peak temperature with warming rate. Activation energies of about 2 and 1.5 eV were obtained for samples with 66 and 5% vinyl concentrations, respectively. These values are in agreement with Alfimov and Nikolskii's own evaluation in accordance with the relation:

$$\frac{1}{T_p} = C_1 - \frac{k}{E} \ln \beta \tag{12}$$

where C_1 is a constant and β is the heating rate. Relation (12) follows from Equation (7) when $E/2kT_p \gg 1$.

Although both Equations (7) and (12) give similar activation energy values, utilization of Equation (7) is preferable because along with the activation energy, it allows evaluation of the frequency factor. According to Partridge's estimate, the frequency factor values

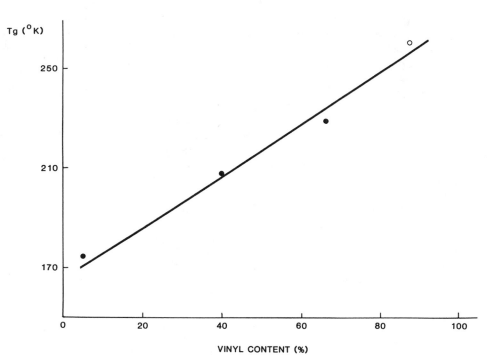

FIGURE 6 T_g dependence on vinyl concentration for polybutadienes.
Solid circles, Ref. 51; open circle, Ref. 52. (After Ref. 40,
Springer-Verlag, with permission.)

are about 10^{45} and 10^{42} sec^{-1} for high and low vinyl content polybutadienes, respectively. These values are very high and exceed
by far a limiting frequency of 10^{12} sec^{-1}, which corresponds to a
transition with zero activation entropy [53]. It should be remem-
bered, however, that we are dealing with the glass transition, i.e.,
with a transition of a complex nature in which a large activation
entropy is expected. Activation entropies of 60 to 100 eu have been
reported for some relaxations in polymers and even in low molecular
weight solids [54]. Thus high-frequency factor values on the order
of 10^{40} sec^{-1} are not unrealistic and first of all indicate a high degree of complexity in the motion associated with the relaxation.
     It is always of interest to compare relaxation parameters obtained
by different methods since this might provide some additional information. In the study conducted by Knappe et al. [55] on cis-polybutadiene (98% 1,4-cis), the activation energy of the glass relaxation
was found to be 1.35 eV by the method of initial rises and only 0.43
eV by the method of various heating rates. Such a discrepancy

cannot be due to quenching (phosphorescence quenching, if taking place, would give just the opposite effect). As a matter of fact, close activation energy values obtained by Nikolskii when the method of initial rises [22] and the various-heating-rates method [51] were applied indicate that phosphorescence quenching is of no importance in this particular case. Since the activation energy in the vicinity of 1.3 eV has also been confirmed by other researchers [57], it is believed that the value of 0.43 eV is underestimated. The most probable reason for this seems to be the insufficient accuracy of the measurements. As was shown by Alfimov and Nikolskii [51], the T_g peak position on the glow curve shifts to higher temperatures with increases in warming rate as expected, but only by about 6° for a warming rate increase of 30 times. In the study performed by Knappe et al. [55], the heating rate varied from 1 to 3°/min, i.e., by three times, which is not adequate to ensure a reliable activation energy evaluation.

B. Polyethylene

First experiments on the thermoluminescence of different polyethylene samples irradiated with γ rays at liquid nitrogen temperature in the absence of oxygen revealed the presence of three low-temperature transitions at the following temperature regions: 140 ± 10, 185 ± 10, and 230 ± 20K [24]. The transition in the vicinity of 140K (the γ peak) was more intense relative to the remainder of the glow curve and much better resolved in high-density than in low-density polyethylene samples. This was considered an indication that the γ relaxation in polyethylenes originates in crystalline regions. The same conclusion has been made on the basis of results obtained for polyethylene samples allowed to stand at room temperature in hexane [59]. The effect of hexane on the polyethylene glow curve resulted in the disappearance of the transitions around 220 and 180K, whereas the lowest temperature transition at 130K (γ transition) was only partially affected by hexane even after prolonged treatment. Partridge and Charlesby [60] found that polyethylene single crystals exhibited very intensive γ relaxation, whereas the luminescence intensity at higher temperatures was small relative to bulk-crystallized material. In a study of the radiothermoluminescence of highly crystalline linear alkanes [61] it was shown that such compounds exhibit a transition in the same temperature region as the γ transition in polyethylenes. Finally, the glow curves for both high-density and low-density polyethylene blown film samples showed a significant decrease in the intensity of the γ relaxation compared to that in bulk-crystallized samples, whereas the latter exhibited a higher degree of crystallinity [15].

The set of experimental facts cited above can be considered as evidence in favor of a crystalline phase origin of γ relaxation in polyethylene. Contrary to the γ transition, complete removal of the transitions located at temperatures above the γ transition as a result of heptane treatment indicated that they originate in polyethylene amorphous regions [59].

Figure 7 presents the radiothermoluminescence glow curves for two low-density polyethylenes [15]. In both cases, three low-temperature transitions are resolved. For both polymers, the temperature position of each low-temperature transition is the same, whereas the intensities are different. Since different polyethylene samples show temperature transitions located at the same temperature intervals, each of the transitions can be assumed to be derived from the mobility of a similar structural form. In this case, the intensity of a particular transition may be considered to be proportional to the mass fraction of a structural formation which is manifested by this transition.

In order to have a better understanding concerning the proper designation of each of the low-temperature transitions to mobility in specific structural regions, it is convenient to compare the data obtained by radiothermoluminescence analysis with the results obtained by other means.

A series of papers presented by Kitamaru and Horii [62—64] showed that three-component nuclear magnetic resonance spectrum

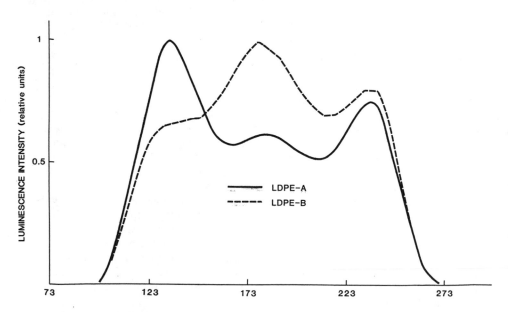

FIGURE 7 Glow curves of two quenched low-density polyethylene samples. (After Ref. 15, John Wiley and Sons, with permission.)

analysis gives detailed information on the multiphase structure of polyethylene samples in terms of the molecular mobility or relaxation modes associated with each phase. It was concluded that bulk crystals generally have a structure composed of crystalline, interfacial, and interzonal materials. These interfacial and interzonal components are associated with limited and liquidlike molecular mobility, respectively. The relative content of these components, as well as the molecular conformation or mobility of each component, varies over a wide range depending strongly on molecular weight and temperature. The temperature dependence of the phase structure for annealed, monodisperse, linear polyethylenes of three different molecular weights as observed by Kitamaru and Horii [64] is shown in the middle Section of Figure 8. For an intermediate molecular weight sample at 120K, the spectrum comprises only the crystalline component with no interfacial and interzonal components. The interfacial component is identified from the spectrum at 170K in addition to the broad component. The narrow component is observable at temperatures above 230K. These results suggest a large temperature dependence of the molecular mobility in each phase. The authors indicated that the appearance of the interfacial component around 170K and the interzonal component around 230K corresponds to the low-temperature relaxation processes in polyethylene. A similar analysis has been made with samples of low and high molecular weights.

The left side of Figure 8 presents the radiothermoluminescence results obtained for annealed monodisperse linear polyethylene samples with molecular weights similar to those of the samples analyzed by Kitamaru and Horii. The difference in molecular weights for high molecular weight samples is not important. The nuclear magnetic resonance technique has shown that samples of any molecular weight higher than 100,000 behave similarly to the sample with a molecular weight of 431,000.

Let us now compare the results obtained by both the nuclear magnetic resonance and radiothermoluminescence techniques. First of all, according to nuclear magnetic resonance, the appearance of interfacial and interzonal mobility shifts to lower temperatures with an increase in polyethylene molecular weight. There is a similar shift in the positions of the two thermoluminescence maxima. The highest temperature transition shifts from 238K for a low molecular weight sample to 232.5K for a high molecular weight sample. The corresponding shift for the intermediate temperature transition is from 198 to 178K. For samples of progressively higher molecular weights, the increase in intensity of the radiothermoluminescence maximum around 230K also correlates with the increase in mass fraction of the interzonal component. Along with these similarities, there is one basic dissimilarity in the results obtained by the two techniques. According to nuclear magnetic resonance, the mass fraction of interfacial material increases with the increase in molecular weight, where-

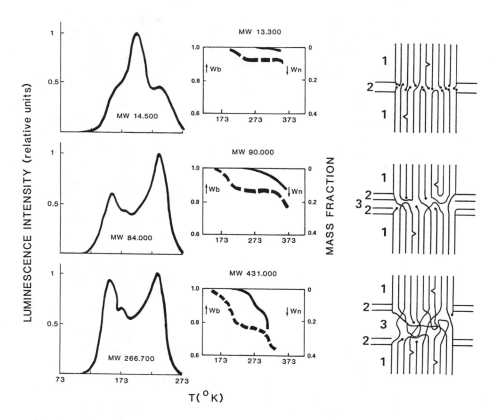

FIGURE 8 (left) Glow curves for linear polyethylenes. (center) Mass fractions of three components measured by the nuclear magnetic resonance technique as a function of temperature for linear polyethylenes. Solid and dashed lines indicate narrow and broad components, respectively. Wn, Wb = mass fractions of narrow and broad components, respectively (see Ref. 64). (right) Schematic structure models of different molecular weight linear polyethylenes: disheveled, unpeeled crystal for low molecular weight sample and lamellar crystals for high and medium molecular weight samples. 1, 2, and 3 indicate the crystalline, interfacial, and interzonal regions, respectively. (After Ref. 15, John Wiley and Sons, with permission.)

as radiothermoluminescence seems to show just the opposite effect, i.e., a decrease in combined intensity of the intermediate transitions located at 178 and 198K. Let us consider this discrepancy in more detail. The spectrum of highly crystalline HNO_3-treated polyethylene at 123K was taken as the elementary spectrum for the crystalline component. Although HNO_3 treatment probably destroyed the external noncrystalline portion of the material, it could not remove or heal the internal crystalline defects. So if some local molecular motion appears in the internal part of the crystalline region at temperatures above 123K, its contribution would be assigned to the interfacial, but not crystalline, component by the nuclear magnetic resonance technique. A somewhat similar situation has been noted by Sinnott [65], who observed a certain level of low-temperature mechanical relaxation in polyethylene samples having no nuclear magnetic resonance mobile fraction. Now, turning back to the radiothermoluminescence analysis, the higher the molecular weight, the greater is the intensity of the transition at 153K (Figure 8). This trend can be easily understood if one takes into account the increase in the degree of crystal imperfection with molecular weight. Thus the dissimilarity between the nuclear magnetic resonance and the radiothermoluminescence data is only an apparent one and can be explained by postulating that all three radiothermoluminescence transitions at 153, 178, and 198K contribute to the interfacial nuclear magnetic resonance component.

The right side of Figure 8 shows schematic structural models which correspond to low, intermediate, and high molecular weight polyethylene samples. These structural models are similar to those described by Kitamaru and Horii [64]. The only difference in our model is the presence of internal crystalline defects. The structure of a disheveled, unpeeled crystal is shown as the analog of the structure of a low molecular weight sample. Medium and high molecular weight samples have lamellar crystal structures. The concentration of defects in the crystalline region increases with molecular weight. The low molecular weight sample has many chain ends and few loops and tie molecules in the interfacial region. The interzonal region consists predominantly of tie molecules and manifests itself to a greater extent in the samples of high molecular weight. Accepting this structural scheme, each of the radiothermoluminescence low-temperature transitions can be attributed to a specific structural region and to structural units in polyethylene, e.g., transition around 140K, crystalline region defects—kinks, jogs, etc.; transition at 178K, interfacial region—cilia, loose loops; transition at 198K, interfacial region—chain ends, tight loops; and transition at 233K, interzonal region—tie molecules. Further, taking into account that the thermoluminescence intensity in the region of each particular transition is proportional to the mass fraction of the structural unit manifested by the transition, one can try to predict the mechanical properties of polyethylene on the basis of the radiothermoluminescence results. In order to achieve this goal, we must try to understand the influence of the different types of structural units on the mechanical properties of polyethylene. Such an estimation is to a large extent approximate; however, several conclusions are apparent (Table 4). For samples of the same degree of crystallinity, I would

TABLE 4 Influence of Different Structural Units in Polyethylene on Its Mechanical Properties

Property	Defects (kinks, jogs) crystalline region	(Cilia, loose loops) interfacial region*	(Chain ends, tight loops) interfacial region*	(Tie molecules) interzonal region*
Tensile at yield	No influence	−	−	+
Elongation at yield	No influence	+	−	−
Impact strength	No influence	−	−	+
Environmental stress crack resistance	No influence	−	−	+

*Positive influence = +; negative influence = −.
Source: Ref. 15.

TABLE 5 Basic Mechanical Properties of Two Low-Density Polyethylene Samples

Property	Sample A	Sample B
Tensile at yield (psi)	1730	1530
Elongation at yield (%)	18	31
Low temperature brittleness— WECO Notched (°C)	-41	+10
Environmental stress crack resistance—ASTM 10% Igepal (hs)	168	0.08

Source: Ref. 15.

like to underline the positive influence of tie molecules and the negative influence of chain ends on the basic mechanical properties of
polyethylene, e.g., yield strength, impact strength, and environmental stress crack resistance. However, crystalline defects seem
not to have any influence, whereas an increase in cilia has a negative
effect on yield strength and environmental stress crack resistance
and a positive effect on elongation. Table 5 shows the basic mechanical properties for two low-density polyethylenes whose glow
curves are presented in Figure 7. Sample A exhibits a higher intensity of the transition at 233K (more correct, a larger area under
the maximum at 233K), which is indicative of a greater concentration
of tie molecules. However, the intensity of the transition at 178K
is higher for sample B. Thus one can conclude that this sample has
a larger cilia concentration. Since tie molecules predominantly have
a positive influence on the basic mechanical properties and cilia have
a negative effect, one should expect better mechanical properties
for Sample A. The experimental results support this conclusion.

C. Polypropylene

As opposed to polyethylene, which exhibits transitions in well-de
fined temperature regions, there are in many cases striking differences in the reported polypropylene transition temperatures even
when the studies were carried out by means of the same technique
[66]. The only transition which was reported by practically all
researchers is the glass transition of amorphous polypropylene that
lies slightly below 273K, although it is uncertain whether there is a
difference between the glass transition in atactic and isotactic materials [67,68]. Different groups of authors found none or several
transitions in the range 173 to 273K, and in some cases transitions
were also observed at lower (down to 19K) and higher (up to 373K)
temperatures. I believe that the essential dissimilarities in polypropylene transition temperatures reported in the literature are due
primarily to a low level of isotacticity and thus to the high atactic
and stereoblock content in some of the polymers introduced as isotactic.

Properties of polypropylene are to a great extent dependent on
tacticity and the amount of low molecular weight soluble material in
the polymer. Fractional extraction with different solvents in a
Soxhlet-type apparatus is widely used in industry for the preliminary
screening and evaluation of material quality.

Semicrystalline polymers such as polypropylene can be considered
to be composite materials, consisting of purely amorphous and purely
crystalline regions, with a wide morphological spectrum in between.
It is expected that the dissolution behavior of such a polymer is complex, with contributions from each region. Above T_s, the thermo-
dynamic dissolution temperature, all fractions of the polymer are

thermodynamically soluble, but in a short time amorphous and low
molecular weight crystalline fractions are kinetically preferred.

Glow curves for three polypropylene samples with the same heptane-insoluble content (89.5%) are shown in Figure 9. Despite the
same level of heptane insolubles, the differences in shape of the glow
curves indicate significant structural differences between the samples. At the same time, one notices that the temperature intervals
where the intensity of emitted light exhibits maxima are similar for
all the samples, namely, 273 to 263, 223 to 213, and 143 to 133K.
In one case (curve c), a poorly pronounced transition around 173K
can also be noticed. The radiothermoluminescence evaluation of a
large number of polypropylene samples produced by different manu-
facturers showed that these same transitions are characteristic for
all polypropylenes having a relatively high level of isotacticity [69].

The so-called atactic fraction obtained from the customary heptane extraction of polypropylene actually may consist of a series of

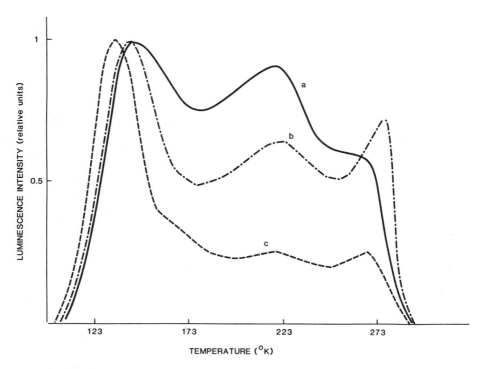

FIGURE 9   Glow curves for three polypropylene samples with the
same heptane-insoluble content.   (After Ref. 69.)

various components in variable weight ratios which are similar only
in their solubility in heptane: noncrystallizable atactic polypropylene, crystallizable relatively low molecular weight stereoblock polypropylene, and extremely low molecular weight isotactic polypropylene.   Thus both heptane-soluble and insoluble fractions can be expected to contain various components depending on the actual composition of the material before extraction.   This aspect has been underscored by several authors [70,71], and it was emphasized that
the evaluations based on material balance only may be misleading.
The radiothermoluminescence results support this conclusion.   Figure 10 presents the glow curves for two virgin polypropylene samples (A and B) and their heptane-soluble and insoluble fractions.
The major difference between samples A and B is the greater intensity of the broad transition centered at 223K and the higher temperature of the transition in the vicinity of 273K for sample A.   At the
same time, the dissimilarities in the glow curves of the heptane-soluble and insoluble fractions of these materials are dramatic.   The soluble fraction of sample B most probably represents completely atactic
polypropylene with a narrow maximum in the T_g region and essentially broader secondary relaxation transition at 213K.   Since the
lowest temperature transition for amorphous polypropylene is located
at 213K, the transitions at 133 and 163K must arise from relaxation
in polypropylene crystalline regions unless the relaxation behavior
of the amorphous phase formed by atactic and isotactic chains differs essentially.   Following this line, it can be concluded that the
soluble fraction of sample A contains some crystallizable material.
This results in increased width of the maximum around T_g and its
shift by 8° to higher temperatures as compared to sample B, and it
also results in the appearance of the two low-temperature transitions
at 133 and 163K.   The characteristic features of the insoluble portion of sample A are an intensive broad transition at 223K, the appearance of a very weak transition at 163K, and a poor resolution
of the thermoluminescence maximum in the region of the glass transition.   The major differences between virgin sample B and its heptane-insoluble fraction is in the position and shape of the most intensive lowest temperature maximum.   For the heptane-insoluble
fraction, the temperature position of this maximum (150K), as well
as its large width, indicates that it may be composed of two overlapping transitions of about the same intensity located at 163 and
133K.
     The limited radiothermoluminescence results on polypropylene re-
laxation obtained so far do not permit a substantiated assignment of
different transitions to specific structural formations.   It has to be
underlined, however, that as with polyethylene, relaxation in polypropylene manifests itself in certain temperature intervals.   An additional indication of this was obtained by evaluation of highly isotactic
polypropylene samples of different molecular weights [69].   The same

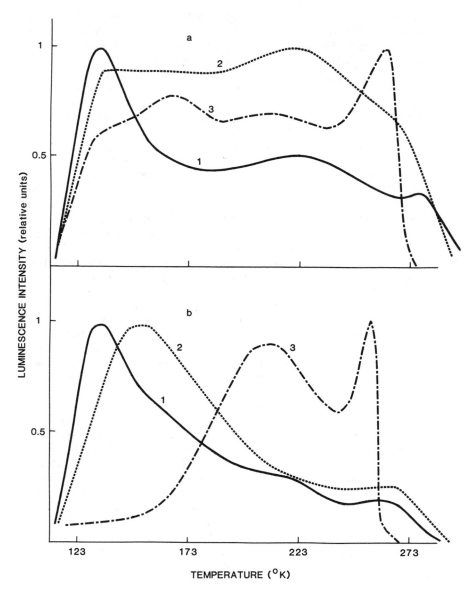

FIGURE 10 Glow curves for two polypropylene samples: (a) sample
A and (b) sample B (1 = virgin materials; 2 = heptane-insoluble
fractions; 3 = heptane-soluble fractions. (After Ref. 69.)

four transitions as mentioned earlier were observed. The intensity of the transitions around 133 and 263K increased with molecular weight, whereas two other transitions at 163 and 223K were barely pronounced, especially in low molecular weight samples.

The only definite conclusion which can be made at the present time is that completely amorphous atactic polypropylene exhibits glass relaxation and secondary relaxation around 260 and 220K, respectively. It is still unclear, however, whether the amorphous phase in isotactic polypropylene can be identified with that in atactic material. The transitions at 163 and 133K most probably are caused by the relaxation of different defective sites in the polypropylene crystalline phase.

It is noteworthy to mention that, as a rule, commercially available atactic polypropylenes are not completely amorphous and exhibit both amorphous and crystalline transitions [72]. Thus any studies on polypropylene relaxation have to be performed on carefully and thoroughly characterized samples.

D. Polymer Blends and Copolymers

The most widely used method for evaluating polymer compatibility is determination of the transition temperatures. Compatible polymers are considered to be those which form a single-phase blend with a single glass transition temperature dependent on the composition of the blend. Polymer blends which exhibit two or more T_g values corresponding to the glass transition temperatures of the individual components are considered to be incompatible.

The transition temperature criterion for compatibility evaluation provides the ground for application of the radiothermoluminescence technique to the study of polymer blends. In the heterogeneous blend comprising two separate phases, the fate of a secondary electron generated on radiation exposure will depend on the size and shape of the domains and on the average travel range of secondary electrons [9]. This can be rationalized by using Figure 11, a schematic representation of a region near the interface between domains of the phase-separated blend components polymers A and B. When both polymer phases are in the glassy state, the secondary electron generated in and traveling through the frozen matrix A will most likely be trapped at T_A in phase A. If the secondary electron is generated near the AB phase boundary, it can, however, cross the interface and be trapped at T_B in phase B. One can make a similar argument for the trapping of secondary electrons generated in phase B.

On subsequent release of the secondary electrons from the traps on warmup, the electrons trapped in phase A should more or less all combine with ions present in the same phase. It will very likely be parent ion $I_A{}^P$ from which the electron originated or another ion

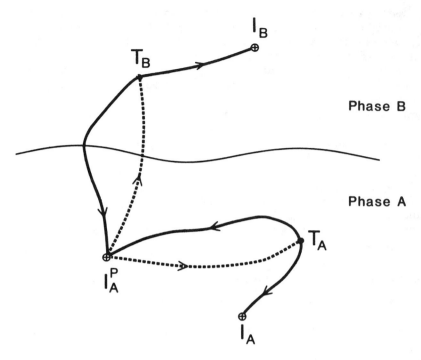

FIGURE 11 Ion recombination process. Schematic drawing of electron trapping near the domain interface in a heterogeneous polymer
blend. (After Ref. 9, 1979, American Chemical Society, with permission.)

I_A in the vicinity of the trapped electron. Those electrons which
transferred and were trapped in phase B could in principle recombine with ions I_B of phase B. Again, however, it is probable that
most of these electrons, attracted by the coulomb field of their parent ions, will return to phase A and recombine there. The reason
for this is that once an electron is released from one of the traps,
its energy is only kT. Without much of its own initiative, its travel
and range are governed by the coulomb field of ions in its vicinity.
The separation between positive and negative ions required to reduce
the energy of coulomb attraction to thermal energy kT at T = 77K
is about 1000 Å. Since the average distance of initial charge separation is estimated to be 60 Å [6] and the average distance between
two ions of about 200 Å [75], we can conclude that the parent ion,
on average, will be the closest neighbor to the trapped electron.
Hence it is likely to recombine with it. Thus it can be assumed
that at T greater than T_g^A and T_g^B, essentially all the electrons

created in phase A will have returned to A and most electrons created in phase B will have returned to B.

For domains much larger than 60 Å—the average travel distance
of the secondary electron following irradiation—practically all electrons generated in each phase will be trapped in this phase and the
glow curve of the blend can be expected to be the sum of the glow
curves A and B. Yet the glow curve of the blend may not be the
sum of A and B. This is so because the electrons crossing the
phase boundary from B to A will do so at a transition temperature
characteristic of polymer B. However, this effect may be pronounced
only when phase domains are not essentially larger than 60 Å. Thus
a deviation of the glow curve of the blend from that of A plus B
can be visualized even in the hypothetical case of small phases divided by geometrical borders. In a real system with diffusive in-
terfaces between domains, the appearance of such deviation will depend not on the phase dimensions but rather on the volume contribution of the interface.

Let us now discuss some of the experimental results. Figure 12
shows the glow curves for cis-polybutadiene—ethylene-propylene copolymer blends of different compositions [19]. The variations in
the ratio of the components result in a gradual increase in entensity
of one of the maxima with a corresponding weakening of the other.
The T_g values determined from the positions of the maxima are 175
and 209K over the whole composition range and coincide with those
of cis-polybutadiene and ethylene-propylene copolymer, respectively.
Moreover, the width of the two main peaks are about equal to those
of the pure materials, hence indicating that no appreciable interface
exists between the domains of the component polymers. The polybutadiene—ethylene-propylene copolymer blend is a typical example
of a heterogeneous system with practically no phase interaction.
Thus the intensity of luminescence originating in separate phases is
proportional to their volume fractions. This provides the linear dependence of the intensity of each of the T_g maxima on the concentration of the components.

Several radiothermoluminescence studies have been performed on
the compatibility of polybutadiene with butadiene-styrene copolymer
[9,18,20,75,76]. This blend provides a rare system in which phase
uniformity can be varied by changing the conditions of preparation
and heat treatment. Figure 13 shows the glow curves of polybutadiene, butadiene-styrene copolymer (styrene/butadiene ratio 20:80),
and a blend comprising 50 wt% of each of the elastomers prior to and
after annealing at 423K [75]. As can be seen, the glow curve of
the blend even prior to annealing cannot be approximated by the
sum of the glow curves of the components. The essential broadening and overlapping of the transition peaks indicate a certain degree
of solubility of the components even at room temperature. After
annealing, instead of the two dominant transitions at 178 and 202K,

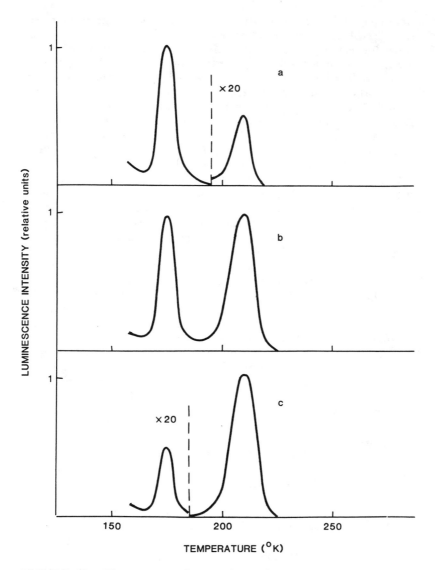

FIGURE 12 Glow curves for various cis-polybutadiene—ethylene-propylene copolymer blends. The weight fractions of cis-polybutadiene in the blends are (a) 0.97, (b) 0.50, and (c) 0.03, respectively. (After Ref. 19, with permission of the Rubber Division of the American Chemical Society.)

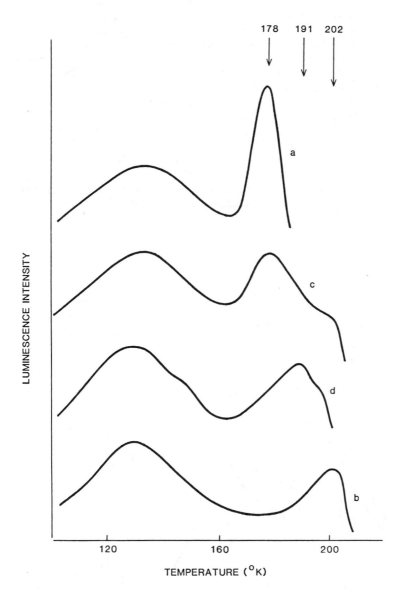

FIGURE 13 Glow curves for (a) polybutadiene (<u>cis</u>/trans/vinyl =
54/35/11) and (b) poly(butadiene-co-styrene) (20% styrene), and a
blend comprising 50 wt% of each of the two elastomers prior to (c)
and after (d) annealing in a vacuum at 423K for 26hr. (After Ref.
75, with permission of the Rubber Division of the American Chemical
Society.)

only one strong peak at 191K can be noted and, on its shoulder, a
less pronounced transition at about 197K.   Thus the compatibility of
the polymers is considerably enhanced by annealing, which leads to
the formation of a pseudohomogeneous system possessing an ensemble
of microregions which differ in composition.   This is characteristic
for a system with a developed interface.

Finally, Figure 14 presents some of the results obtained for the
blend of cis-polybutadiene with high-vinyl polybutadiene [28,77].
In this case, the glow curves of blends of different compositions are
characterized by a single maximum situated between the maxima of
the individual polymers.   Similar glow curves were obtained independently of the method of preparation (roll-mill mixing or solvent evapora-
tion).   Neither quenching of the blends to liquid nitrogen temperature
nor prolonged annealing at 423K changed the shape of the curves.

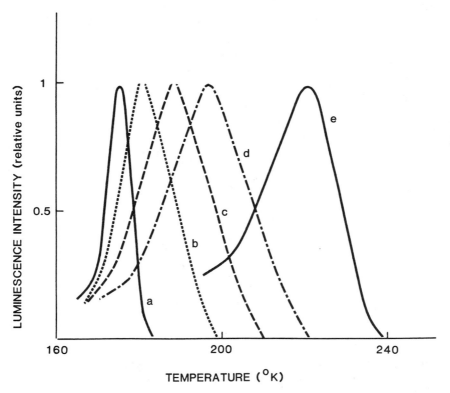

FIGURE 14   Glow curves for (a) cis-polybutadiene and (e) polybutadiene with 60% vinyl content and their blends with cis-polybutadiene
weight fractions of (b) 0.7, (c) 0.5, and (d) 0.3.   (After Ref. 77,
Pergamon Journals, Inc., with permission.)

In accordance with the glass transition temperature criterion of compatibility, it can be concluded that cis-polybutadiene and high-vinyl polybutadiene are compatible in a whole range of concentrations.

Relaxational behavior of copolymers resembles that of polymer blends and depends on the length of the individual sequences. Copolymers with long-chain block or graft sequences of different natures are equivalent to heterogeneous polymer blends, i.e., they exhibit two glass transitions coinciding with those of the individual components. Alternatively, when block or graft lengths are decreased to a level approaching the molecular structure of a random copolymer, only one glass transition located on the temperature scale between the glass transitions of the components is observed. In an intermediate case of short blocks and, consequently, small phase dimensions, the influence of the interface is significant, resulting in a broadening and shifting of the glass transitions toward each other.

E. Latex Systems

As distinguished from solid rubbers or plastics, latexes are colloidal suspensions of polymer particles in water. Upon being dried, a film-forming latex is transformed from a milky dispersion into a transparent film in which the contours of the particles observed initially gradually become less distinct.

Because intermolecular interaction is lower at the surface than in the bulk [50], the opportunity for relaxation is much greater at the surface than in the interior of a particle [56]. Thus differences in relaxational behavior have to be expected between latex- and solution-cast films. I am not aware, however, of any methods except radiothermoluminescence [58] which succeed in clearly demonstrating this.

Figure 15 shows the glow curves of natural rubber films prepared from latex and a solution in benzene. Both samples have a distinct T_g maximum of the same width at 212.5K. At the same time, the luminescence peak at 183K is much more intensive and better resolved in the latex film. This is most probably linked to the presence of a developed interface between the latex globules on which the kinetic mobility of polymer segments is higher than in the interior. The intensity of the peak at 183K increases with decreasing latex particle size and decreases with the duration of aging, which leads to a gradual disappearance of the interface, or with the addition of a stabilizing mixture (electrolyte solution), which promotes aggregation of the globules.

The final technological step in the processing of both solid rubbers and rubber latexes is vulcanization. There are, however, certain differences between the two cases caused by the peculiar globular structure of latex. Additives introduced into the latex in the form of an aqueous dispersion do not distribute themselves uniformly between the serum and the concentrate in the interglobular spaces, such as are created during the formation of a rubber film from la-

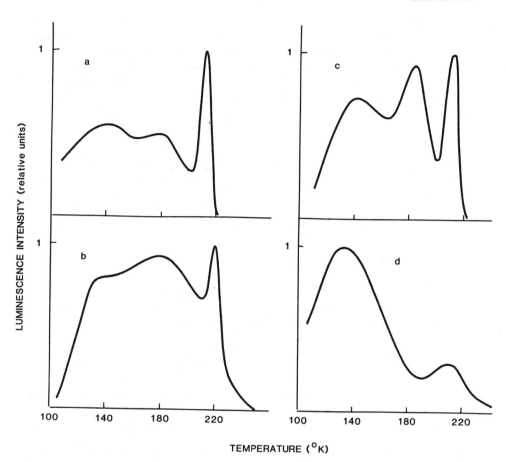

FIGURE 15 Glow curves for (a, c) unvulcanized and (b, d) vulcanized natural rubber films cast from (a, b) solution and from (c, d) natural latex. M_c of vulcanized films was 5500. Composition of the vulcanizing agents (parts per 100 parts of polymer by weight): sulfur, 2.0; ZnO, 3.0; and ethylcymat, 1.0. Vulcanization was performed in an air thermostat at 200°C. (After Ref. 58, John Wiley and Sons, with permission.)

texes [73]. Two processes take place during the warmup at vulcanization temperature: (1) vulcanization of a rubber layer on the globule surface which is in direct contact with the vulcanization system, and (2) diffusion of the vulcanization system from the surface layer into the polymer particle. These processes determine the initial rate of vulcanization, which is therefore higher in a latex (due to higher vulcanization system concentration in the interglobular space) than in a solid rubber.

Vulcanization of the natural rubber film prepared from solution does not change the shape of the glow curve substantially (Figure 15). The peak around 183K remains. The position of the peak in the T_g region is shifted to higher temperatures, and it becomes wider, which is typical for vulcanizates of hard elastomers. Cross-linking, by restricting certain types of motion, causes a spread in the distribution of relaxation times characterizing the process. A glow peak thereby covers a greater temperature range, the spread being in the high-temperature direction. Differently from the solution film, the glow curve of the latex film changes considerably after vulcanization. The maximum at 183K disappears. Unfreezing of segmental mobility in the T_g region begins at lower temperatures in the vulcanized film than in the nonvulcanized film. The luminescence maximum in the T_g region broadens considerably without changing its position along the temperature axis.

These distinctions are apparently connected with the different structures of films prepared from solution and latex films. The homogeneous structure of films from the solution, as opposed to latex films, as well as the more uniform distribution of vulcanization agents in them creates conditions for a relatively identical bonding of the polymer throughout the entire volume. Hence the vulcanization of films from solution does not substantially affect their phase homogeneity. Typical of a latex film, however, is a globular structure with definite interstices between the globules. The vulcanizing ingredients are concentrated in these interstices in the immediate vicinity of the mobile segments on the globule surface. This, in turn, creates conditions under which vulcanization affects mainly the surface layers of the globules and thus leads to intensive interglobular bonding. As a result, the free interglobular surface decreases and the relaxation transition at 183K vanishes. Nevertheless, a somewhat greater mobility of the outer globular layers even after crosslinking is apparent in the glow curves from the fact that the increased segmental mobility in the T_g region begins at lower temperatures in vulcanized latex film than in nonvulcanized film. An abnormal broadening of the maximum at T_g, owing to the vulcanization of latex film, results directly from its microheterogeneity and nonstatistical distribution of crosslinks.

F. Oriented Systems

Orientation of semicrystalline polymers results in a broadening and a shift of the thermoluminescence maxima toward high temperatures [74]. In the region of Hookean deformation, the shift is relatively small—5 to 7°. It sharply increases when yield stress is reached and then levels off in the region of forced rubberlike elasticity. Finally, it increases again at large strains, i.e., when extension is due to extension of straightened macromolecules. The magnitude of the shift in transitions depends not only on the extent of deformation, but also on the temperature at which the material was strained [78]. At low temperatures, where relaxation is suppressed, orientation and, consequently, the upward shift of the transition temperatures increase with an increase in temperature. At high temperatures, relaxation is a determinant factor, and the opposite effect is observed. Thus the shift in transitions is defined not by the absolute value of deformation but by the level of unrelaxed mechanical stress accumulated in the sample. For polyethylene, the temperature of deformation which gives the largest shift in transition temperatures was found to be 300K [78].

Interesting results were obtained when, instead of orientation and subsequent irradiation at liquid nitrogen temperature according to the standard radiothermoluminescence procedure, these two operations were switched around [79]. The original polyethylene film exhibited two major thermoluminescence maxima at 181 and 232K, which arise from interfacial and interzonal mobilities, respectively, and can

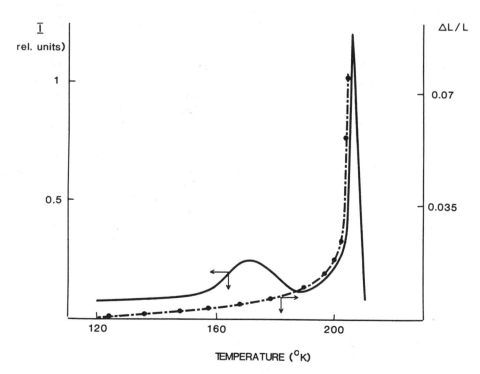

FIGURE 16   The glow curve for low-density polyethylene film under 400 kg/cm² stress. The dashed line shows the change in relative elongation with temperature. (After Ref. 79.)

be identified in terms proposed by Boyer [14] with $T_g(L)$ and $T_g(U)$ transitions. When the stress applied to the preirradiated polyethylene film was greater than 270 kg/cm^2, there was a new flash on the glow curve (called the δ peak by the authors) which appeared at the moment of film rupture (Figure 16). For stress less than 270 kg/cm^2, the δ peak was not observed and the rupture occurred when luminescence practically vanished. At this condition, a shift to lower temperatures of both thermoluminescence maxima could be noted (Figure 17 a, b). For stresses exceeding 600 kg/cm^2, the rupture was accompanied by a negligible elongation and the temperature of the δ peak reached a constant value (~170K), which lies in the region of $T_g(L)$ relaxation and can probably be identified with the polyethylene brittleness temperature (Figure 17c). This indicates the connection between the onset of ductility and the polyethylene $T_g(L)$ relaxation.

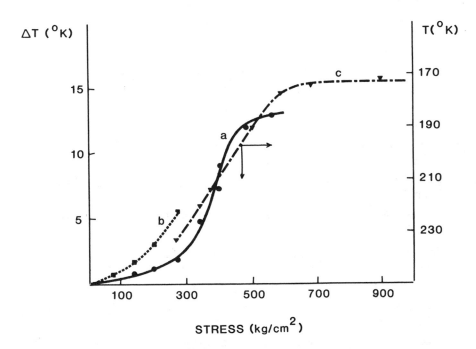

FIGURE 17 Shift in the positions of the transitions (a) $T_g(L)$ and (b) $T_g(U)$ and (c) the position of the δ maximum as a function of applied stress. (After Ref. 79.)

VIII. CONCLUSIONS

The radiothermoluminescence technique is simple, rapid, and very sensitive to the onset of various types of molecular motion in organic materials. Thus, it is an excellent method for the evaluation of glass and secondary transitions in polymers. The method is also well suited for studies of polymer blends and copolymers, oriented films and fibers, and latex systems as well as structural changes accompanying polymer crystallization, crosslinking, and plasticization.

Radiothermoluminescence is especially suitable for evaluating temperature transitions at low temperatures, although there are several reports stating the usefulness of the technique at temperatures essentially above room temperature. Possible high-temperature applications provide an interesting field for future studies.

The most serious drawback to radiothermoluminescence as it has been conducted in the past was the use of a large source of ionizing radiation. The recent development of the scanning electron microscope attachment containing the cryogenic stage and the vacuum feedthrough manipulator promises to make the radiothermoluminescence method more convenient and widespread. *Graphs.*

REFERENCES

1. R. H. Partridge, The Radiation Chemistry of Macromolecules, Vol. 1 (M. Dole, ed.), Academic Press, New York (1972).

2. G. N. Lewis and M. Kasha, J. Am. Chem. Soc., 66: 2100 (1944).

3. M. V. Alfimov, V. G. Nikolskii, and N. Ya. Buben, Kinet. Catal., 5: 238 (1964).

4. A. Charlesby and R. H. Partridge, Proc. Roy. Soc., A271: 188 (1963).

5. I. Boustead and A. Charlesby, Proc. Roy. Soc., A315: 271 (1970).

6. F. Williams, The Radiation Chemistry of Macromolecules, Vol. 1 (M. Dole, ed.), Academic Press, New York (1972).

7. M. Dole and D. M. Bodily, Adv. Chem. Ser., 66: 31 (1967).

8. A. Ekstrom, R. Suenram, and J. E. Willard, J. Phys. Chem., 74: 1888 (1970).

9. G. G. A. Bohm and K. R. Lucas, Adv. Chem. Ser., 174: 227 (1979).

10. V. G. Nikolskii, V. A. Tochin, and N. Ya. Buben, Sov. Phys. Solid State, 5: 1636 (1964).

11. V. G. Nikolskii and N. Ya. Buben, Proc. Acad. Sci. USSR Phys. Chem., 134: 827 (1960).

12. M. Magat, J. Chimie Phys., 63: 142 (1966).

13. N. N. Semenov, Pure Appl. Chem., 5: 353 (1962).

14. R. F. Boyer, Macromolecules, 6: 288 (1973).

15. L. Yu. Zlatkevich and N. T. Crabb, J. Polym. Sci., Polym. Phys. Ed., 19: 1177 (1981).

16. Kh. G. Mindiyarov, L. Yu. Zelenev, and G. M. Bartenev, Polym. Sci. USSR, A14: 2347 (1972).

17. V. G. Nikolskii, Pure Appl. Chem., 54: 493 (1982).

18. V. G. Nikolskii, L. Yu. Zlatkevich, V. A. Borisov, and M. Ya. Kaplunov, J. Polym. Sci., Polym. Phys. Ed., 12: 1259 (1974).

19. L. Yu. Zlatkevich and V. G. Nikolskii, Rubber Chem. Technol., 46: 1210 (1973).

20. L. Yu. Zlatkevich, Rubber Chem. Technol., 49: 179 (1976).

21. V. G. Nikolskii, L. Yu. Zlatkevich, M. B. Konstantinopolskaya, L. A. Osintseva, and V. A. Sokolskii, J. Polym. Sci., Polym. Phys. Ed., 12: 1267 (1974).

22. V. G. Nikolskii and G. I. Burkov, High Energy Chem., 5: 373 (1971).

23. V. G. Nikolskii, Sov. Sci. Rev., 3: 77 (1972).

24. A. Charlesby and R. H. Partridge, Proc. Roy. Soc., A271: 170 (1963).

25. R. J. Fleming, J. Polym. Sci., A2, 6, 1283 (1968).

26. L. F. Pender and R. J. Fleming, J. Phys. [C], 10: 1571 (1977).

27. G. G. A. Bohm, J. Polym. Sci., Polym. Phys. Ed., 14: 437 (1976).

28. N. Ya. Buben, V. I. Goldanskii, L. Yu. Zlatkevich, V. G. Nikolskii, and V. G. Raevskii, Polym. Sci. USSR, A9: 2575 (1967).

29. L. Zlatkevich, L. F. Nichols, and N. T. Crabb, J. Appl. Polym. Sci., 25: 963 (1980).

30. A. Linkens and J. Vanderschueren, J. Electrostat., 3: 149 (1977).

31. A. Linkens and J. Vanderschueren, J. Polym. Sci., [B], 15: 41 (1977).

32. I. K. Lednev, V. A. Aulov, and N. F. Bakeev, Proc. Acad. Sci. USSR Phys. Chem., 265: 659 (1982).

33. G. C. Meggitt and A. Charlesby, Radiat. Phys. Chem., 13: 45 (1979).

34. R. H. Partridge, J. Polym. Sci. [A], 3: 2817 (1965).

35. P. L. Land, J. Phys. Chem. Solids, 30: 1681 (1969).

36. P. N. Keating, Proc. Phys. Soc., 78: 1408 (1961).

37. V. Maxia, S. Onnis, and A. Rucci, J. Lumin., 3: 378 (1971).

38. S. Onnis and A. Rucci, J. Lumin., 6: 404 (1973).

39. G. F. J. Garlick and A. F. Gibson, Proc. Phys. Soc., 60: 574 (1948).

40. L. Zlatkevich, Radiothermoluminescence and Transitions in Polymers, Springer-Verlag, New York (1987).

41. H. Gobrecht and D. Hofman, J. Phys. Chem. Solids, 27: 509 (1966).

42. A. Bohun, Czech J. Phys., 4: 91 (1954).

43. I. A. Parfianovitch, J. Exp. Theor. Phys. USSR, 26: 696 (1954).

44. W. Hoogenstraaten, Philips Res. Rep., 13: 515 (1958).

45. A. G. Wintle, J. Mater. Sci., 9: 2059 (1974).

46. T. Hashimoto, T. Sakai, and M. Iguchi, J. Phys., [D], 12: 1567 (1979).

47. A. E. Blake, A. Charlesby, and K. J. Randle, J. Polym. Sci. Polym. Lett. Ed., 11: 165 (1973).

48. N. Ya. Buben, V. D. Grishin, V. G. Nikolskii, V. L. Talroze, and V. A. Tochin, Br. Patent 1285650 (1972).

49. L. Zlatkevich and B. Crist (in press).

50. V. E. Vettegren and A. Chmel, Vysokomol. Soed. [B], 18: 521 (1976).

51. M. V. Alfimov and V. G. Nikolskii, Polym. Sci. USSR, 5: 477 (1963).

52. G. G. A. Bohm, J. Polym. Sci., Polym. Phys. Ed., 14: 437 (1976).

53. H. W. Starkweather, Macromolecules, 14: 1277 (1981).

54. W. Kauzmann, Rev. Mod. Phys., 14: 12 (1942).

55. W. Knappe, G. Voigt, and A. Zyball, Colloid Polym. Sci., 252: 673 (1974).

56. R. F. Boyer, Polymer, 17: 996 (1976).

57. S. M. Aharoni, J. Appl. Polym. Sci., 16: 3275 (1972).

58. L. Zlatkevich and M. Shepelev, J. Polym. Sci., Polym. Phys. Ed., 16: 427 (1978).

59. I. Boustead and A. Charlesby, Proc. Roy. Soc., A316: 291 (1970).

60. R. H. Partridge and A. Charlesby, J. Polym. Sci., Polym. Lett. Ed., 1: 439 (1963).

61. I. Boustead, Proc. Roy. Soc., A319: 237 (1970).

62. R. Kitamaru, F. Horii, and S. H. Hyon, J. Polym. Sci., Polym. Phys. Ed., 15: 821 (1977).

63. R. Kitamaru, F. Horii, and S. H. Hyon, Polym. Preprints, 17: 546 (1976).

64. R. Kitamaru and F. Horii, Adv. Polym. Sci., 26: 137 (1978).

65. K. M. Sinnott, J. Polym. Sci. Polym. Symp. [C[, 14: 141 (1966).

66. A. E. Woodward and J. A. Sawer, Physics and Chemistry of the Organic Solid State, Vol. 2 (D. Fox, ed.), Interscience, New York (1965).

67. F. E. Karasz and W. J. MacKnight, Macromolecules, 1: 537 (1968).

68. D. R. Burfield and Y. Doi, Macromolecules, 16: 702 (1983).

69. L. Zlatkevich (unpublished results).

70. A. Nakajima and H. Fujiwara, Bull. Chem. Soc. Japan, 37: 909 (1964).

71. A. Nishioka, Y. Koike, M. Owaki, T. Naraba, and Y. Kato, J. Phys. Soc. Japan, 15: 416 (1960).

72. I. Boustead and T. J. George, J. Polym. Sci., Polym. Phys. Ed., 10: 2101 (1972).

73. V. Duchacek, Int. Polym. Sci. Technol., 9: T/35 (1982).

74. A. D. Shulyak, V. S. Yerofeyev, et al., Polym. Sci. USSR, 13: 1234 (1971).

75. G. G. A. Bohm, K. R. Lucas, and W. G. Mayes, Rubber Chem. Technol., 50: 714 (1977).

76. S. S. Pestov, V. N. Kuleznev, and V. A. Shershnev, Kolloid-Z., 40: 581 (1978).

77. N. Ya. Buben, V. I. Goldanskii, L. Y. Zlatkevich, V. G. Nikolskii, and V. G. Raevskii, Proc. Acad. Sci. USSR Phys. Chem., 162: 386 (1965).

78.   V. A. Tochin, D. N. Saposhnikov, and V. G. Nikolskii,
      Vysokomol. Soed. [B], 12: 609 (1970).

79.   V. G. Nikolskii, D. N. Saposhnikov, and V. A. Tochin,
      Vysokomol. Soed. [B], 12: 19 (1970).

Index